# THE UNIVERSE AND ITS LIFE

*A Journey through Space, Time and Mind*

Alfred A. Maske

ATHENA PRESS
MIAMI  LONDON

THE UNIVERSE AND ITS LIFE
*A Journey through Space, Time and Mind*
Copyright © Alfred A. Maske 2002

All Rights Reserved

No part of this book may be reproduced in any form
by photocopying or by any electronic or mechanical means,
including information storage or retrieval systems,
without permission in writing from both the copyright
owner and the publisher of this book.

ISBN 1 931456 13 5

First Published 2002 by
ATHENA PRESS
Queen's House, 2 Holly Road
Twickenham TW1 4EG
United Kingdom

Printed for Athena Press

# THE UNIVERSE AND ITS LIFE

*A Journey through Space, Time and Mind*

*To the Reader
who seeks
Knowledge of Self and God*

Any man who seeks wisdom, wisdom will find him. It will not be hidden from him.

*Aramaic Testament of Levi*
of the Dead Sea Scrolls.

The formulation of a problem is often more essential than its solution, which may be merely a matter of mathematical or experimental skill. To raise new questions, new possibilities, to regard old problems from a new angle, requires creative imagination and marks real advance in science.

Albert Einstein in
*The Evolution of Physics*

# Contents

| | | |
|---|---|---:|
| Introduction | | xi |
| I | Our True Self | 19 |
| II | Our Home, the Milky Way | 35 |
| III | Facts of the Universe | 38 |
| IV | Our True Origin | 42 |
| V | An Eternal Job | 59 |
| VI | Reincarnation: Fact or Myth? | 70 |
| VII | The Stuff of Mind and Consciousness | 87 |
| | About Prophecy | 105 |
| VIII | Spaceship Earth | 111 |
| | Antediluvian Civilizations | 120 |
| | The Geology | 125 |
| | Insights | 130 |
| IX | A New Theory of the Universe | 133 |
| | A New Approach | 137 |
| | Quasars and Antimatter | 148 |
| | Creation Through Annihilation | 152 |
| | Spirit and Matter | 155 |
| | Space Travel and Extraterrestrials | 157 |
| X | A Theory of Everything Or "Quantum-Relativity" | 166 |

| | |
|---|---|
| The Grand Picture | 171 |
| Energies and Forces | 182 |
| **Conclusion** | **235** |
| The Cosmos | 241 |
| The Universe | 242 |
| Our Galaxy (and all others) | 243 |
| The Earth | 244 |
| Gravity | 246 |
| Light Propagation | 248 |
| **Appendix I** | **252** |
| **Appendix II** | **254** |
| **Appendix III** | **257** |
| **Appendix IV** | **259** |
| **Appendix V** | **262** |
| The Ultimate Question | 262 |
| **Suggested Reading** | **265** |
| Astronomy | 265 |
| General Knowledge | 265 |
| Human History | 265 |
| Inspirational | 266 |
| Philosophy | 266 |
| Physics | 266 |
| Reincarnation and Karma | 266 |
| Religion | 267 |
| Spirituality | 267 |

# Illustrations

| | |
|---|---:|
| Our Threefold Nature | 61 |
| The Human Brain | 92 |
| The Shielding at Earth and Moon (principle) | 190 |

# INTRODUCTION

> True fortitude of understanding consists in not letting what we know to be embarrassed by what we do not know.
>
> Ralph Waldo Emerson

Why do we exist? What is the purpose of our existence? Is this one the only life we have, or is there more to it? Are we somehow connected to the universe, and why does the universe exist? Has time a beginning and an end? Is space infinite? Do we have a soul, and if so, what is it? Is there a causal intelligence, or God, or does everything exist without a cause? Where and how did we originate? How did millions of life forms appear on our planet? How does the universe work? *What* are space, time, light, infinity, gravity, electricity, magnetism, and all the other basics of science?

If these questions are also yours, you have found the right book. These questions are as old as humanity. Religions throughout the history of mankind have tried to find answers, but all they could offer were certain belief systems or dogmas. Scientists have not done any better. They have had to accept the fact that established facts and beliefs are wrong. Just think about astronomy. 2,000 years ago the earth was flat. Then the sun rotated around the earth for 1,500 years. Next, all planets orbited the sun in perfect circles until Kepler stated: No, they are ellipses! At that time our solar system was the universe. Later on, scientists learned that our solar system is only a tiny member of the Milky Way galaxy, but all the other galaxies were still considered to be nebulae. Only at the beginning of the twentieth century our better telescopes changed the nebulae into what they are: other galaxies exactly as ours.

From time to time, wise man – who have some of the most important answers for the problems of their time – walk the earth. From some we have their wisdom in writing, but unfortunately dozens of different religions were founded after they passed away, resulting in misinterpretations that led to dislike, hate, and even war between the different groups. Just think about the fighting that is going on between Catholics and Protestants in Northern Ireland, or of the so-called holy wars in the Islamic world. The inquisition of the Middle Ages murdered thousands of people in the name of Jesus.

This madness was only caused by the ignorance of people who did not realize that all religions were basically teaching the same – love the other person, do not kill, keep peace and do not start wars, honor your parents – only in different languages and for different mentalities.

Most of us live with these questions. Some do not question very much, and too many people do not care at all. You have these issues on your mind; otherwise you would not have read up to here, or reached for this book in the first place.

These questions and problems have bothered me throughout my life. For more than thirty years, I worked my way through hundreds of different sources of knowledge. While studying, I compared, combined and listed all the results. Finally, I added my own knowledge, logical reasoning, and many epiphanies – the sudden intuitive flashes of understanding that arrive from the subconscious mind. The result is a good collection of answers to all of the above questions and problems. Then I realized that many people would really like to do the same, but do not have the time or opportunity to do so. Just one single book, not too big, but containing almost everything in a condensed form, appeared to be the right thing to do for these people.

Albert Einstein wrote a book titled *The World As I See It*. My book is concerned with the same topic. The difference is that Einstein's book was very specialized from his viewpoint as a physicist, but this book has a broad spectrum over all categories of knowledge. It contains theories, phenomena, and facts of physics, astronomy, biology, space travel, history, prophecy, religions, and metaphysics, all combined into a very big picture. Yes, knowledge of all of these categories is required for a full understanding of the universe, ourselves, and the meaning of life. Spirit and science will be joined together into a total science, covering everything.

Of course, I cannot claim to have the full truth; nobody can. That would be an outrageous claim if we consider the size of the universe and the infinite number of questions it has in store for us. But I believe that this book contains some good guidelines that can help you develop your own opinions. I am also aware that this book will challenge some scientific theories and also some religious dogmas, because in the last chapters you will find my own theories about the nature of the universe, the speed of light, and gravity. They are based on common sense and logic, and can be easily comprehended, unlike the ideas of some scientists of today.

Many scientists are too narrowly specialized, and some are even arrogant about their own fields and do not accept the work and ideas of other scientists. They forgot, or never knew, that the word "science" simply means knowledge. Everything in physics, mechanics, mathematics, philosophy and metaphysics is science, because it is knowledge. Logic, common sense and reasoning are as valuable as laboratory tests for scientific investigation.

Albert Einstein believed firmly that real scientific progress has to begin with an idea and a theory, with the formulation of a problem. The mathematics and experimental skill will be last. I fully agree with his viewpoint. I just cannot agree with the methods of some scientists, who first build a complex mathematical picture and then expect to get proof for it through observation.

Beginning a few decades before World War I, scientists were pressured by the governments to work on the development of new weapons. As a result, the thinking of scientists got more and more materialistic. Now we are told that everything in the universe exists by chance, including life.

Only what can be tested and physically proven is real. Philosophy, logic, god, spirit, soul – these and many other words are taboo and not considered scientific.

But fortunately, many scientists do not believe in these narrow-minded dogmas anymore and have begun to accept the idea that there is something else besides materialism. Of course, they have to be very careful about expressing their convictions or they will lose their jobs or be branded for not being "scientific".

There is one thing that I have to mention here. While reading this book, you will find that you have to think on a very large scale. You have to stretch your imagination to the limits; but it is fun! When you arrive at the end of this book it will be clear that this book introduced you to the *largest possible picture* of the cosmos and of ourselves.

In a way, this book is a piece of mental research and, therefore, I have had to apply a certain method. As I have already mentioned, scientists always want proof for everything new, and they are right to demand it. Otherwise it is not science. But *what is* scientific proof? Only the actual finding of an object, a verified photo, the result of a test, or mathematical "proof" are accepted. Yes, that is the way of scientific thinking today, because there is no way to test and prove the existence of an intangible, metaphysical entity like a soul or spirit. Nor can we take a yardstick and check the voltage in an outlet. Spirit is not testable, and that is why most scientists stay away from non-physical topics.

Of course, some non-physical phenomena manifest sometimes in a way that can be photographed or measured, but most people have never heard about it, and some do not want to hear about it. There are many photographs of spiritual apparitions. Most are of so-called ghosts, but the best-documented photographs are of the apparition of the Virgin Mary on the top of a church in Zeitoun, a suburb of Cairo in Egypt. It started on April 2, 1968 and went on for two years, almost every night, witnessed by thousands of people, most of them Moslems. As a matter of fact, two Moslem mechanics were the first to see the apparition on the roof of the Coptic Church known as St. Mary's Church of Zeitoun. Hundreds of photographs were taken, mostly by professional photographers.

The only instrument we have to test and prove non-physical things is the mind. Besides clairvoyance and intuition, the most effective way of using the mind as a research instrument is logical reasoning. Logic, reasoning, and common sense will be the basic principle for the guideline throughout this book. It is a method that worked very good in ancient Greece, so why not for us?

Basically, we have three groups of people separated by their belief systems. One group is entirely materialistically oriented, believing that everything in the world, including thought and feeling, even emotions, can be explained as a function of matter. The non-scientists of this group also believe that scientists are always right, and they accept everything new from them as a proven fact, even if the whole thing is only a hypothesis – like the Big Bang theory – or silly or exotic. Most of the people of this group think that dealing with metaphysics, religion, or spiritual phenomena is a waste of time, and they are not even willing to investigate into it.

The second group is strictly spiritual-religious oriented. They believe only in their religion or dogma, and nothing else has a chance. Many are very dogmatic and willing to fight about the meaning of a word in their scripture, and they do not see how silly that is. Just one example: in the Bible, Psalm 37:11, we find the statement that the *meek* shall inherit the earth. Many people just cannot agree on what that means. Does it mean the weak, spiritless characters, or are the meek the gentle people without anger? Both meanings exist, but is this important? No! The general message of the scripture is what is important. But worst of all is the fact that so many people are convinced that their own religion is the only right one, all others are judged to be wrong. If we consider the hundreds of different religions, plain logic will tell us that such thinking makes no sense.

And then there is the third group, which fortunately is the largest I think. They are the people who tolerate the beliefs and theories of the other groups. This book is written especially for them. The scientifically oriented reader will find what he wants to know about metaphysics and spirit; the religious-oriented reader will receive scientific information that will help him to understand his religion better.

But this leading together of science and religion is not the only purpose of this book. The main purpose is to introduce you to my own theories about problems in physics, astronomy, geology, and biology. Some of them will be of help, so I hope, to unite the many theories we have today, and to eliminate some theories that just cannot survive logical investigation. The first half of this book will deal mainly with our spiritual heritage and who we are; the second half concentrates on my theories.

In each case we will first analyze the currently accepted theories, and then my own theories will be explained. We will talk about evolution, Albert Einstein's Special and General Relativity, Max Planck's Quantum Theory, Werner Heisenberg's Quantum Mechanic, and of course, the Big Bang theory.

For the reader who is not scientifically inclined, I will treat all these scientific topics as simply as possible, but I must also be accurate enough for the scientist who reads it. Therefore, please do not be scared. It will be very interesting and even fun.

After leaving the dead end road of the Big Bang theory and the presentation of a better way to interpret the red shift observations of Edwin Hubble and all the other problems of cosmology, I will work on the dream of all scientists – the Theory of Everything, or TOE. So far nobody succeeded with TOE, even Einstein left it unsolved on his desk at the time of his death.

But the main purpose and the most important aspect of this book is my hope that it will be of great help for you. My goal is to replace faith or belief with knowledge whenever possible. At least, I hope that I will introduce you to an interesting and fascinating way of thinking and reasoning. I promise: it is fun!

# Chapter I
# OUR TRUE SELF

> Science without religion is lame, religion without science is blind.
>
> Albert Einstein

Before we try to find out what the whole world is about, first we should know *what we are*. What are we? This question is not as easy to answer as it seems. Of course, for some people it is easy. They will tell you: "My body! That is me! It is as simple as that, and my mind is a function of my brain, period." What these people are saying is that they are nothing more than a biological machine; when they die it is all over. But most people have their doubts, deep in their subconscious minds. The fact that we do have a subconscious mind should tell us that we are more than just a biological machine.

There are other hints that point to something else. We know that very *strange things are happening* on this planet every minute, and they are of a non-physical nature. These millions of cases are so overwhelming, if only by their number, that we cannot ignore them. They force us to think and to reason about it. The greater the number of similar cases, the greater the probability will be that we are dealing with a fact.

Let us look at a very common and well-known example: Throughout the history of mankind we have had millions of instances when a person died, and at the exact same minute, over distances of hundreds or thousands of miles, the mother or another loved one woke up and knew that this other person was dead. As we all know, this happens especially and very often during wars.

If we think and reason about this we know for sure that we

must all be interconnected by some unknown, non-physical force that can bridge vast distances in an instant. And even if we do not know what it is, we know at least that we are not just a biological machine and that we must be more than that.

In order to find out *what* we are, we will have to probe into a few more of such "strange" examples. For the purpose of proving my point I will use only well-known cases, nothing entirely new, and then we will reason again.

A widely known phenomenon is NDE or near-death experience. If you have already read Dr. Raymond Moody's book *Life After Life*, then you know the topic. This book is only one among a long line of books about this and related topics, but he started the whole thing, and so his book is considered to be a "classic" in this field. In his work as a medical doctor he witnessed a case and became curious about it. He coined the term "NDE". He started a research program that resulted in this book.

For the reader who is not familiar with the phenomenon of NDE, I will explain. What happens – and this is a proven fact – is that hundreds of people who are pronounced clinically dead for a few minutes and then come back to life have vivid accounts of their experiences while dead. There have been such accounts throughout human history. Thanks to modern medicine, people are now much more often pulled back from death and consequently, there is a high frequency of such cases today. The stories recounted by these people are never the same, but there are so many similarities that we must accept them as being true, and we can "design" an average case. Please note that not all NDEs are pleasant. Almost all people who tried to commit suicide had hell-like experiences. However, a typical story of a patient could go like this:

> Sometime during my operation I regained consciousness, even though I was under anesthesia, and I found myself floating at the ceiling. Looking down, I saw my body on the operating table, and the doctors had just found out that they lost me. Then the scene became wild. More people came running and more machines

were brought to the tableside. The doctors worked very fast on my body and tried to resuscitate me.

Then, and I think that is strange, I lost interest in my body and floated right through the wall into an adjacent surgery room and saw what was going on there. I thought of my wife and instantly I was in the waiting room and saw her sitting in a chair. I tried to contact her, but couldn't get her attention.

Now the picture changed and I floated through a kind of tunnel. Coming out at the other end I saw a bright light in front of me. It was brighter than the sun, but not blinding, just a warm feeling. I could not see anybody, but I knew that somebody was in that light.

A voice asked me if I was satisfied with the life which I had lived on earth. Then, like in a movie, my whole life was shown to me by the apparition of life. It showed every detail of my life, even every thought I had, in just one second, or so it seemed, but I comprehended everything. Somehow I had the impression of being in a world without time. The voice informed me that I would have to go back to my life on earth. I did not want to go back, I never felt so good, and it felt like home. Then suddenly I lost consciousness.

When I came out of the anesthesia, the doctor told me that I had been gone for a few minutes during surgery. I told him that I knew, and explained to him all the details of what I had seen him doing and what I had seen in the other room behind the wall. He verified everything.

Later, when my wife came to my bedside, I told her about my "visit" and described every detail of the waiting room which I had never seen before. She was shocked because all the details were right.

The people who have *these* experiences are very sure that they are real. And all of them are transformed by them into a new state of mind. Now they comprehend things they could not grasp before. I am the best example I can offer you. After my NDE, I needed almost exactly the famous two years for the complete change. Among the many publications which deal with this phenomenon, P. M. H. Atwater explains this "ripening period" in the best way in her book *Beyond The Light*. She is a leading authority in this field. Most of my own theories in this book are a direct result of this aftereffect of my

NDE.

Many so-called experts explain the whole thing away as hallucinations, or as functions of the dying brain. If you ask them how a dying brain can observe everything in a room, leave the room and observe happenings far away, they will have no answer or will change the topic. One psychologist managed to prove with electronical means that a subject experienced the "illusion" of an out-of-body experience (OBE). For him, these physiologic measures were proof that OBEs are all in the brain. What he actually did was to cause an OBE by electronical means, not realizing that thousands of people can do the same with meditation. The experience of his subject was a real one!

These so-called out-of-body experiences, OBEs, are very different from NDEs. Some people have them while meditating, others while in a trauma shock, and some are able to do it at any time they want to. Many succeed in doing it during yoga. We had OBEs in my own family, so I have first hand experience with these phenomena. When you are out of your body during an OBE, you can go to any place you want to visit, even out to space, and you are free to do whatever you want to do, which is not the case with a NDE. That is the big difference between NDEs and OBEs. You do not want the NDE, it happens to you, but you can initiate an OBE. That is why you are free during an OBE to do what you want to do. Another way to see it would be in this way: during a NDE your body is dead, but during an OBE your body is alive.

Because another force imposes the experience of a NDE on you, you have no influence on it. The only thing that you may subconsciously affect is a kind of "coloring" in accordance with your religious beliefs. A Christian, for example, may report seeing Jesus, a Moslem may imagine the presence of Allah, an atheist may wonder what is going on.

And now we are ready for some logical reasoning. Please note that this is *my* reasoning, I did not see it anywhere else.

When we leave the body in an OBE or NDE, we leave the

brain of our body behind; we do not have it with us! But while out of the body, we are thinking, having emotions, and we collect impressions. We can also see and hear while our eyes and ears are still in our body.

Then, when we are back in our body, we have full memory of everything that we experienced while out of the body. What is going on here? Do we have a double nature?

It is the *real selves* that left the body! We even have a name for it. It is the *soul*! Because we can memorize everything under all conditions, in or out of body, our *main memory* cannot be located in our brain, and neither can most of our thinking, reasoning, and emotions. This main "brain" of our real selves, of our spiritual form, is superior to the body brain by magnitudes! True, the brain of our body can perform great and amazing tasks of its own, but it is permanently connected to the spiritual system, because its capacity is limited. (Most of our brain is used for controlling the body, and for the function and interpretation of our five senses, for example, the whole language capability.) In Chapter VII, the whole structure of the brain and mind will be explained in detail. Here it is only required to point out that our scientists are misled when they find out that certain mental functions happen in certain parts of the brain. They do not *happen* there. What scientists locate is the area where certain functions are concentrated and then transmitted to the spiritual system of the real self.

There is an exception, though. Every function of our body is entirely the work of our brain, and has to be learned. But the brain cannot give signals, for example, to the leg muscles without a command from the spiritual system to do so. *We* want to walk, and the brain is the servant and does the physical work.

What is this real self that left our physical body and took our mental life with it, while leaving the brain behind? Most people who studied this field call it the "astral body". They call it astral, because the astral world, or better realm, is the lowest level of the spiritual realm; it is where we go first when we "die". I will use this term too, but please remember it is only a name. The astral body is just another body that we control with our soul by transferring our consciousness into it, but the soul is still much more than the physical and the astral bodies combined, much more! The physical body and the brain is just a temporary possession, required for dealing with the physical world of matter. It is *not* our real self.

To summarize, the brain is a high-performance computer that controls our body, processes all impressions we receive through our senses, and then, functioning as a sophisticated switch system, transfers everything into the "astral brain". The other way around, it receives all orders, ideas, etc. which are generated in our (soul) mind, from the astral brain, and the result is bodily movement or action.

This is not easy to conceive because we tend to identify ourselves with the body we are born with.

There is another interesting fact that surfaced in most of these experiences: *love*! Love seems to be not just an effect, or a manifestation of feelings; it clearly appears to be a *force*! It is the universal force that rules the whole spiritual realm and radiates down from the astral realm into our physical world.

In almost every NDE, when the light-being appeared to be a critic and told the soul that the past life was not a good one and that the entity had failed in many ways, this was always done with a lot of understanding and helping love. It was a judgement, but in the form of a helpful critic – even for entities who had been a bad person in life.

But people who tried to kill themselves, or some people who did something *very wrong*, did not receive this treatment of love; instead they received a hell-like treatment. Most of the time they arrived in total darkness after their body died. Such a shock treatment brings everyone down to an absolute nothing. But when the soul finally understands that it did something very wrong and cries for mercy, the darkness will be taken away and this entity will experience the light and the love of the light-being like all the others did right away. A soul cannot be destroyed! This topic, love as a form of energy, will be clarified in detail in a later chapter.

Of course, some scientists just do not believe that a soul or any kind of non-physical condition exists. They had to make tests in the laboratory, and as could be expected, with negative results. But there is one test that proves that love is a force. I read it years ago in a book, but I cannot recall what book it was. Anyway, this is the test:

> Two subjects, a man and a woman who were deeply in love, were placed in different rooms and later even several miles apart. Each had a gadget placed on the head which measures the intensity of brain waves – a machine very similar to an electrocardiogram (ECG) for the heart, but sensitive to the much weaker electrical signals of the brain. Both places had the timing of the brainwave recorders synchronized precisely with each other.
>
> And then came the "fun". At an exactly predetermined second, without the knowledge of the two subjects, one of the two got stung with a needle or was otherwise mildly hurt. As expected, the recorder for the person being hurt showed a peak at this instant. And then the so-called surprise: the other person in this "love connection" had the same peak on the recorder at exactly the same second.

Is it really a surprise as it was called in that book? I do not think so. I believe that this test just proved in another way that all of us are interconnected. Earlier in this chapter we have already talked about people suddenly waking up at the very moment a loved one dies. It is the same thing.

All this observed facts and the test results prove two things with certainty: first, that love *is* a powerful form of energy, and second, that our mind can perform things that are not physical in nature. An unknown, invisible force or energy seems to be at work. But what is it?

The best and most fully documented proof of our spiritual nature and capabilities were the telepathic powers of a man who lived in the first half of this century in the United States. (He had other spiritual abilities too.) His name was Edgar Cayce. He performed more than 15,000 clairvoyant "readings", as they were called. They are safely kept in a fireproof vault at the Association for Research and Enlightenment (ARE) in Virginia Beach, VA, very well organized and cross-referenced for every classification. Before I get to the real reason for mentioning Edgar Cayce in this chapter, I must explain who he was and what he did.

Edgar Cayce was a very simple man. He could not finish high school because he had to work on his father's farm. What did he do? He diagnosed the illnesses of thousands of people without even seeing them and in most cases over distances of hundreds of miles. Then he gave instructions for treatment. He did many other astonishing things that we will talk about later.

There was no crystal ball, incense, or a darkened room. In broad daylight he simply lay down on a couch, and within two minutes he put himself into a deep trance-like state. At a certain sign that he was ready, the question for the case at hand was given to him. Medical histories were never given to him, only the fact that a person needs helps, his name and address; nothing more! His lifelong secretary, Gladys Davis, took down in shorthand everything he said. The people who were seeking help were sometimes present in the room, but most of the time they were hundreds or thousands of miles away.

After a while Cayce would say, "Yes, we have the body." (He always said "we". It is something to think about.) Then followed a complete diagnosis, with all the fancy medical terms that Edgar Cayce did not know anything about when awake. After that he suggested what should be done for a cure of the illness. Then he would contact the ailing person by mail or telephone.

If the person in question, or the doctor, followed his advice, the patient was cured or at least got much better. But some did not follow his advice because it seemed too strange to them. He had 100 per cent success if we count only the cases where people followed his advice. The others just do not count; the readings for them were futile.

In 1923, just by coincidence, he began a long series of so-called life readings, which dealt with ancient earth history, prophecy, and reincarnation.

Of course, as always, there were many people who could not believe that Edgar Cayce was for real. One New York businessman put Cayce to the test by asking him to trace all his steps and doings on the way to his office in a reading. On that day, the businessman did everything completely different from his usual routine. For example, he used the stairs instead of the elevator. At exactly the same time, Edgar Cayce was on his couch in Virginia Beach and gave a reading about what the man did in New York.

The executive was flabbergasted when he received the report of the reading in the mail. Edgar Cayce had described not only every step and move in perfect detail, he had even read the letters from the man's morning mail!

It is not the purpose of this book to prove Edgar Cayce's authenticity. This has already been established by the ARE (The Edgar Cayce Foundation), and by hundreds of books about his work. As a matter of fact, he was tested again and again, year after year, so that we can honestly say that he is better verified than, for example, Albert Einstein, from whom we only have words, ideas and formulas.

Most scientists who do research on the function of the mind go strictly by the assumption that the mind is nothing more than a chemo-electrical function of the brain, and here Edgar Cayce comes in!

The following reasoning will show why I had to mention Edgar Cayce for this chapter: what he did is absolutely impossible for a brain to do in any physical or electrical way! A brain cannot go back and forward in time. A brain, located in Virginia Beach, cannot scan a man's body in a hospital elevator in Switzerland, and give a diagnosis that doctors could not arrive at. (The doctors had given up.) Then he gave advice for treatment which worked because the doctors were smart enough to follow his strange advice. A brain cannot do that over a few thousand miles across the ocean but Edgar Cayce did! Not his body, not his brain, but his highly developed entity; the spiritual part of Edgar Cayce, the real him, did the job!

I would like to relate one more example of Edgar Cayce's work which demonstrates many facts to think about. It is kind of scary for the uninitiated.

Once in his trance-like state in Virginia Beach, he had to give a reading for a woman in Chicago. All he got was the name and the address at Lincoln Street in Chicago. No hint whatsoever about the problems of the woman. After a while he said, "Yes, we have the body... Very nice red morning robe." (She wore it so that Cayce could find her better, which was not necessary.) Then he started a complete "physical" of the whole body, finding nothing wrong with her health. After a short pause he mentioned that she had a very bad abscess on her right leg. Then he advised, "Do not amputate!" (Her doctor was planning to do just that.) He then explained that the bite of a poisonous brown spider was the cause. He reiterated that amputating the leg was not necessary. Instead he suggested the use of "Oil of Smoke". Then he gave the name and address of a pharmacy in Chicago that had this medicine.

Cayce, then, called the women in Chicago and reported the result of the reading to her. Somebody went to the pharmacy but returned empty-handed. The pharmacist, who owned the pharmacy for only a few years, knew that "Oil of Smoke" was an old patent medicine long out of production. He checked all his shelves without success. Then he called Edgar Cayce in Virginia Beach, who immediately gave another check reading. The result was then forwarded by telephone to the pharmacist. Cayce said that the pharmacist would find the "Oil of Smoke" on the left side of a certain shelf *behind* a big dark-blue bottle. The "Oil of Smoke" was behind it! This old patent medicine cured the women's leg, and amputation was not necessary.

If we think about this one case, one out of thousands, then we have to recognize the fact that many things go on in this world that we cannot sense, comprehend, or measure in any physical way. Some part of us can leave our bodies and travel instantly over thousands of miles. Arriving at the destination, this mysterious part of us can do almost everything. It can contact the soul of a person (who knows everything about its body) and receive a complete diagnosis of this body. Or, knowing only the name of a product, it will find out where it can be found in the world.

Because these facts cannot be measured in any physical way, we have to apply *logical reasoning*. It is the only way to find the truth, and we have to accept clear results of reasoning as proof.

The following anecdote illustrates the extent of interconnections within the non-physical realm. Let us think about the following Edgar Cayce reading: for a person who had problems with the thyroid, Edgar Cayce "prescribed" *atomidine*, an iodine compound. Nobody had ever heard of it. Finally, after more check readings and much research, the Cayce team learned that *atomidine* was still under development in India, and not on the market in the U.S. They managed to get a sample. How did the entity of Edgar Cayce know about this? The only logical answer I can think of is that Edgar Cayce and the rest of us are not only spiritual entities, connected to each other, but that there exists a huge spiritual recording system about everything, and Edgar Cayce was able to probe into it.

I think we are now ready to answer the question of *what* we are. We are a spiritual entity, or being, which has consciousness and is in control of a physical body, a biological machine we call our body! Most of us also identify ourselves with this body because that is where our consciousness is located while we own this body. And now we know that it is an illusion if we state that this body is the only thing there is. The universe has many aspects of reality, and the physical world is just one of them. We are much, much more than only thinking animals running around on this planet which we call our home. We are not only carbohydrate, protein, fat and minerals; we are much more!

After finding out *what* we are, it follows that we should know *who* we are. This question is not as easy to answer with documented cases. Now we enter an area where tests, physics or mathematics are of little help. But as I have already said a few times, we have another very good method to close in on the truth: logical reasoning and common sense! We have to select a few reliable sources and learn what they know. Then when we compare them, we find great similarity among these sources.

When I was a boy, I overheard a discussion between my father and one of his friends. If I remember right, it focused on the topic of faith and belief. When the friend mentioned to my father that he believes only what he can see with his own eyes, my father asked, "What do you think is inside your skull?" His friend answered, "My brain, of course!" On that my father replied, "Did you see it?"

Here he used exactly the kind of reasoning I have in mind, the same kind I have already explained earlier in this book: if there are millions of cases where a hidden fact was found to be always the same, without exception, we have the right to assume that we will find the same if we probe into it. Logic is a valuable tool for research.

This kind of reasoning can be applied if we investigate a problem that cannot be proven in a test. If thousands of people wrote or said the same thing, independent of each other, and over a period of thousands of years, then we can assume that we are dealing with a fact. I used the word "independent" for a reason. If a dogmatic institution, the church for example, forces the people into the belief that the sun rotates around the earth, then the false belief of the masses is not independent, and we cannot use it for our purpose.

And now to our question of who we are. Edgar Cayce was completely right in thousands of readings. Therefore, by the same logic, we can consider him to be a reliable source. And what did he say about who we are?

It is so simple that it can be expressed in one sentence: God created us, our souls, out of his own spirit and gave each of us a free will and our own consciousness; by definition He made us an *individualized spirit*. Common sense tells us that therefore we are in reality a part of God!

May I point out that Edgar Cayce was a devoted Christian? He read his bible from cover to cover once for every year of his life. This fact influenced the vocabulary of his readings to a high degree. Instead of cosmic intelligence, universal mind, or causal being, he always used the terms *God* or the *Lord*. But all of these names are only man-made terms for a force we cannot comprehend, and therefore are of no great importance. (The Vedas, the ancient sacred books of Hinduism, say that if you cannot comprehend Him, you cannot name Him.) I am also accustomed to the word *God* and so are most people in the Western world. So I will use this term when appropriate. But please keep in mind that the word connotes a personality or at least a personal force. Later on we will see that this is not so. He is not male or female.

He is the universal mind, the causal intelligence, and His body, if we want to call it such, is *everything* in the universe. He caused and He is every atom, every galaxy, all energies and forces, all our souls, and every thought in all history. He *is* everything, and He is *in* everything. This will be explained in detail and from another viewpoint in Chapter X.

Edgar Cayce was not the only man known for the pronouncement of our essential nature; there was another man who taught and wrote about the question of who we are. Judging from what he not only said, but did, I think we can judge him also as a reliable source.

In the days before his death on March 7, 1952, in Los Angeles, he told his followers that he did not want to have his body embalmed when he dies. He also hinted that he would die soon, but was not understood by his followers because he talked euphemistically about an appointment. (He had absolutely no health problems that could lead to death.)

Documents from the Forest Lawn Memorial Park in Glendale, California, reveal that his body was not only fresh and lifelike during the first minutes after his death, but remained in this state without any sign of decay for twenty days in an open casket. There was no smell, no disintegration whatsoever. At the end of this statement the mortuary director stated that the case of Paramahansa Yogananda was absolutely unique in the history of the mortuary business, worldwide.

This man, Paramahansa Yogananda, did not just simply die. He did what true masters have to do. He demonstrated that he had full control over his body after death, just as Jesus and many others did.

On the day of his "death" he was a guest of honor at a special function in the Biltmore Hotel in Los Angeles. Hundreds of people, including the Mayor of Los Angeles, were in attendance. His last picture, taken minutes before his death, was with the wife of the Ambassador of India. After this he gave a short speech, ending with the final lines of a poem: "My India!" With these last words he fell down. Doctors were called and pronounced him dead. No cause for his death could be determined.

Because he told his followers that he had an "important appointment" that day, they now knew that Yogananda died of his own free will, doing what every great yoga master of his kind does. He left his body in a final *mahasamadhi*! (This is a fully conscious exit from the physical body, never losing consciousness during this transition.)

What did he do here in the United States? His Guru, Swami Sri Yukteswar, told him to go to the States and teach the law of love and the similarities between Eastern and Western religions, which he accomplished with great success. His main work, the book *Autobiography of a Yogi*, provides examples that illustrate that each of us is a spiritual entity, a soul, a being who is part of God, and possessing a free will and consciousness. (In India, he was called "the incarnation of love", while his guru, Yukteswar, was "the incarnation of wisdom".) In essence, he echoed Edgar Cayce about the question of who we are.

How do these yogis gain such tremendous wisdom and knowledge? It is with their special kind of yoga, the so-called Kriya Yoga, which is considered to be a scientific kind of yoga. Step by step they turn their body functions off, so that their soul can exit their bodies for thirty to sixty seconds. It is dangerous (It is *not* an OBE), because they even stop their heart for these seconds and are, in a way, dead for this time. During that time (a NDE) they are in a realm without time, and they have a concrete plan of what they want to accomplish, unlike the OBEs of the average person. What one can learn in these eternal seconds is practically without limitation. (Please do not try this yourself without experienced guidance! You may not come back after these seconds.)

Even the bible contains writings about this kind of deep meditation. Many biblical people knew and mastered the technique of this kind of yoga, or deep meditation. Many people believe that Paul in his First Epistle to the Corinthians revealed that he knows about it too. 1 Cor. 15:31 reads as follows: "I protest by your rejoicing which I have in Christ Jesus our Lord, *I die daily.*" What could he mean by dying daily, if not this kind of yoga or deep meditation?

The readings of Edgar Cayce say that John the Divine was also a master of this technique. He handed down a complete instruction for this technique in a coded form. It is the Revelation. It is coded in such a way that nobody can add his own thoughts when translating, it forces every translator to translate literally. Why? Because it can only be understood by spiritual means! And that is what Edgar Cayce did. In twenty-four discourses (Long readings with check readings.) he explained the Revelation as what it is: an instruction for meditation. The entire study is available as the book *Revelation* from the ARE. It is a great surprise for everyone who believed that the Revelation is about prophecy. In a way John provides a hint about the content of the revelation, when he begins Rev. 1:10 with the words "I was in the Spirit…"

There are many more reliable sources that answer the question of who we are, but I think the examples of this chapter should be sufficient for one to arrive at a sure answer to this question.

We are individualized spirit. We have our own free will and we have consciousness. We are a living soul. We identify ourselves always with the body in which our consciousness is located. At this time we are incarnated in a physical body here on earth, and therefore identify ourselves with this body – what an illusion! Each one of us is a part of the causal intelligence of the cosmos, the universal mind, or God. We are like a cell in the body of God, so to speak. He does not allow any cell to die, and if a cell is sick it must be healed. We are created in His image, which means that we are a spiritual entity with free will, consciousness, intelligence, and the desire to create. And, naturally, we did exist as long as He has, and we will exist as long as He will: for eternity!

Here I would like to give a warning. As nice as all of this sounds, we must be very careful with our thoughts. The causal cosmic intelligence, who is everything and who is in everything, causes the physical cosmos to run through His thought energy, which is the most powerful energy form there is. Being part of Him, our thoughts have the same potential. We not only become what we eat, we also become what we think! Later chapters will make that clear.

As long as our consciousness is locked up in our physical bodies, we cannot comprehend the source of our existence, the great causal intelligence, God. But having intelligence, at least we can guess and reason. This capability is our attribute of being a part of God, and we should use it.

# Chapter II
# OUR HOME, THE MILKY WAY

> There are more things in heaven and earth, Horatio, than are dreamed of in our philosophy.
>
> Hamlet

Before we can continue with the investigation of our spiritual heritage, this and the next chapter are required for a better understanding of the rest of the book. So let us talk astronomy, but please do not be scared. I will make it as simple and as short as possible.

Of course, everybody knows that we live on planet earth, but the earth is not our only home, as we will see later. First, I would like to bring our physical home, planet earth, into perspective. To do so, we must have the right measuring stick. The best for our purpose, and the one most commonly used is the speed of light, which is 300,000 kilometers per second (or 186,400 miles per second). Therefore the distance of one light second is 300,000 kilometers. (The science of the world, including the U.S.A. works only in metric. So from here on I will use the original metric data most of the time.)

To illustrate the speed of light, and to get a feeling of it, we can try to imagine it in this way: In one second, light would go around the earth at the equator 7.5 times. That is fast! One light minute would be 18 million kilometers, and a light hour is 1.08 billion kilometers. But as large as they appear, none of these units is used in astronomy, except within our solar system. They are so small as to be useless. Astronomy works with light years, the distance the light travels in one year: 9.46 trillion kilometers. This is a number that is hard to comprehend. (Of course, congress works with such numbers, but nobody there seems to comprehend them. If they did, they would not waste our money by the billions.)

These measures tell us where the earth is located. While we may think of our planet as the center of the universe, we are but a very small part of a huge galaxy, the Milky Way.

We are 150 million kilometers away from the sun and the light needs "only" about eight minutes to reach us. The solar system has eight more planets, with Pluto being the one that is farthest away, 6 billion kilometers away from the sun, and the light requires more than five hours to travel all the way. The order of planets, from the sun out to the furthest reaches of our solar system, is as follows: Mercury, Venus, Earth, Mars are the so-called inner earth-like planets, then follow Jupiter, Saturn, Uranus and Neptune, the outer giant gas-planets. The last one in our solar system is tiny Pluto. We do not know if Pluto is ice or rock; it is too far away.

Our sun has a diameter of 1.4 million kilometers, 110 times the diameter of the earth. One million three hundred thousand earths would fit into the sun. But as big as it appears to us, our solar system is a very tiny unit. To find out how tiny it is we have to go far out into space.

Our sun is a member of the so-called local group, which consists of the close neighbor suns (or stars). Alpha Centauri, Sirius, Tau Ceti and Procyon are the best known of them. They are from 2½ to 15 light years away, very "near".

The suns of our local group are then part of a total of at least 100 billion stars, which make up our "home galaxy" – the Milky Way. This spiral galaxy has a diameter of 100,000 light years, and our sun is approximately 32,000 light years from the center, close to the outer edge. We rotate around the center of the galaxy once every 230 million years at a speed of 220 kilometers per second. Our rockets are snails in comparison.

Our Milky Way galaxy is a typical, average size spiral galaxy; 100,000 light years in diameter, with a maximum thickness of only 13,000 light years at the ellipsoidal bulge, or nucleus, at the center. The spiral arms are half as thick.

Around the whole spiral system is a so-called halo of almost spherical shape and much more than 100,000 light years in diameter. The halo is sparsely populated by globular clusters, each containing up to a million stars. The halo contains also the electrons from the solar winds of 100 billion stars.

Because of our location at the inner edge of our spiral arm, we can observe a very high percentage of the universe with our telescopes. (The center of the arms contains many nebulas.) Our solar system is located in the Orion Arm, named after the constellation Orion. Its stars are only a few hundred light years away, within our arm of the galaxy.

Our sun is not even an average star among the over 100 billion suns of our galaxy, it is only a so-called yellow dwarf. Most stars are larger, many are giants, and some are super giants of a size beyond imagination. How big? Our own yellow dwarf seems to be very big for us with its diameter of 1.4 million kilometers, 110 times the diameter of the earth. But Betelgeuse, in the constellation Orion, has a diameter of 700 million kilometers; 800 times the diameter of the sun! That is almost the diameter of the Jupiter orbit (778 million kilometers). Mercury, Venus, Earth and Mars would all be inside of this giant.

We cannot see the center of our galaxy because interstellar clouds of dust block the view at optical wavelengths, but radio telescopes and X-ray observations provide us with some information. The best ideas of how the center of our galaxy may look come from the observation of other galaxies like ours. The stars in any galaxy are closer towards the center and so closely packed that we cannot distinguish between them. The center appears to be one bright mass. In the very center a "black hole" is expected, which will be explained later.

The next chapter will show why I call the Milky Way galaxy "our home", and not the earth. Our earth amounts to no more then an extremely tiny speck of dust within the universe. Our galaxy represents more than a tiny ball of dust.

# Chapter III
# FACTS OF THE UNIVERSE

> Time and space are the elements of man's own concept of the infinite, and are not realities as would be any bodily element of the earth...
>
> Edgar Cayce

How does our Milky Way galaxy measure up within the universe? Even less than our sun does within our galaxy. We can say that solar systems are the atoms of a galaxy, and galaxies are the molecules of the universe. They represent the basic building blocks of the universe.

Our Milky Way galaxy is a member of another local group. It is called the local group of galaxies. Andromeda and M33 are the best known because they are also spiral galaxies like ours. (There are also elliptical and spherical galaxies.) This local group is a cluster of some twenty galaxies, loosely bound together by gravity. The diameter of the local group of galaxies is 4 million light years.

This is by far not the end; even the next larger group is local. It is called the local supercluster and contains hundreds of galaxy clusters, including our own group. The galaxy clusters within the local supercluster range in size from less than ten, to thousands of galaxies like the Virgo cluster, which is the center of our local supercluster. We are approximately 45 million light years away from the center of Virgo. The diameter of our supercluster is 150 million light years. Superclusters are the largest celestial formations.

Of course, our supercluster is not the whole universe, but we already talked about millions of galaxies just within one supercluster – and every galaxy contains an average of

300 billion suns! Already we get the impression that even a sun seems to be not very important.

And now for our last step into the known universe. By "known" I mean within the reach of our telescopes. The universe contains thousands of superclusters, most as big as ours, but also a large number of solitary galaxies.

Now we can summarize what we know about the universe: There are thousands of superclusters, each containing millions of galaxies with about 300 billion suns in each one. This huge universe, which nobody can fully comprehend anymore, is surrounded by millions of quasars at the farthest rim. Each quasar emits an amount of energy that is equal to the output of many galaxies, but quasars are not larger than a solar system. So far it is anybody's guess what quasars really are. (My own theory about quasars will appear in Chapter IX.) The diameter of the universe is estimated to be approximately 40 billion light years, measured at the far out rim where the last quasars are. (Other estimates say it is only 30 billion light years, which shows how difficult it is to measure these distances.) It is estimated that the universe contains a total of about 100 billion galaxies, which translates into more suns than there are grains of sand on our beaches.

Now we know why so many astronomers talk about the earth as a speck of dust in the universe. But for us it is still very big, very important, and it is our home!

I hope that I have painted a somewhat clear picture of the position of our earth within the universe. We live on the surface of a tiny ball of matter, which is so unimportant that it makes no difference if it ceases to exist. But I intend make you feel better by what we shall see in later chapters in this book: Each one of us, not the body but the soul, is much more important than the earth, and even more important than anything physical in this universe. A soul cannot be destroyed, and any attempt to do so will fail.

If a planet, a sun or a galaxy is destroyed by any means, the matter, atoms, and subatomic particles will remain, some of it transformed into energy. Nothing gets lost. But if a soul could be destroyed, then a cell would be missing in the "body" of God. As such, this is not only a no-no, but absolutely impossible! We are immortal spiritual entities.

In Genesis 1:26, we are given, among other things, "dominion over the earth and what is on it" which, by logic, also includes responsibility for the earth. And the words "Let us make man in our image", also in Gen. 1:26, does not mean in the image of an old, wise man, but as a spiritual being. The physical body was designed later (Gen. 2:7). The whole book of Genesis has to be taken with much caution anyway. Moses was a highly educated man and perfect in his use of Egyptian writing. But the fellow, who translated the books of Moses into Hebrew, a few hundred years later, was not competent to do it. His knowledge of Egyptian was mediocre. He did a lot of guessing instead of proper translating.

Of course, the last paragraph reflected only what I remember from the many so-called "New Age" books I read. The topic of how much of the Bible we can take literally is a very puzzling one. Most of the Bible texts were originally written in Aramaic, the language of the people of the Old Testament and Jesus. Then the texts were translated into Latin and Greek, and the first errors were made, because Aramaic has a small vocabulary compared to Latin and Greek. Later on, translations into English and German were performed. Again a lot of guesswork was required because these two languages are very different from Latin and Greek.

But in this century George M. Lamsa learned that the Church of the East, Middle East to India, still had the Peshitta, the authorized Bible that is still unchanged in Aramaic. These original writings were not available to the translators of the King James Bible and to Martin Luther at their time. Lamsa translated directly from the original ancient Aramaic manuscript, and that is the Lamsa Bible we can buy in any bookstore.

This is the Bible I go by for my own use, but in this book I will stay with the King James Version because that is the Bible most Christian readers have. Please note that the difference between these two is tremendous.

You may wonder what all of this has to do with the universe and astronomy. In Genesis we have the Bible Story of the creation of the world, and these last paragraphs have the only purpose, to tell you to be cautious when you read it. And in the rest of this book we will touch upon the words of the Bible a few more times.

I think that this chapter provided a good picture of our place in the universe. But why we live on this violent and unstable planet and not somewhere else in the universe will be disclosed in Chapter V. All of us have some not-so-good reasons to be on this unsafe, plague, crime and war-ridden planet. But first we have to find out where we came from.

We have now the right perspective to think about a very strange fact here on our earth. Looking from outer space at this unimportant, but beautiful blue-white sphere, we will be bothered by the question: why do the people of this planet behave as they do? Too many of us act not just crazy, but outright stupid. Why? They take the smallest differences or problems so seriously that fights and wars begin. From out there in space, looking at this tiny blue-white ball in an infinite space, such behavior clearly appears to be what it is – stupidity! And we can also see the reason for this condition from the viewpoint of this perspective: it is the counterforce of love – ignorance.

Why so many people are ignorant of the real conditions of their lives has to do with the question: "Why are we here on earth?" The answer to this question, as I see it, will be very different from but better than the stories of all the religions here on earth, because it makes sense and you can feel in your heart that this is the truth. In Chapter V you will find a full explanation of why the answer is so different from current beliefs.

# Chapter IV
# OUR TRUE ORIGIN

> I cannot believe for a moment that life in the first instance originated on this insignificant little ball which we call Earth... The parts which combined to evolve living creatures on this planet of ours probably come from some other body elsewhere in the universe.
>
> Thomas A. Edison

Now we have arrived at the famous question: What came first, the chicken or the egg? The answer is that there was a chicken and a rooster, then came the first egg.

At this time we have two theories about the question of where our bodies came from. And, of course, they completely contradict each other. One is the story of creation, as taught by many religions and the other is evolution, a theory based on the writings of Darwin. Neither one can be proven, but as strange as it may sound, there are many ways to disprove them. So far we have only these two theories. For some strange reason, no third theory has been proposed so far.

We have talked about the book of Genesis, the creation story of the Bible and about Moses and the bad translation of his writings. The man who did this translation work from Egyptian into Hebrew did not have the so-called Rosetta Stone, which was found in Rosetta, Egypt, at his disposal. It is a basalt tablet, now in the British Museum, which is inscribed with a decree from Ptolemy V of 196 B.C. The Rosetta Stone has three complete alphabets side by side: Greek, Egyptian hieroglyphics and demotic characters. It provided the key to the deciphering of hieroglyphs. The French genius Jean Francois Champollion managed to transcribe this tablet into modern language, and since that time we have been able to decipher everything written in Egyptian hieroglyphics.

Genesis still contains a general line of basic truth, but the details comprise, to say the least, a strange puzzle full of paradoxes. The other religions have nothing better to offer. None of our religions can account for the many facts we know about through the diverse branches of science. For example we have found ruins of lost civilizations. Some are dated more than 100,000 years old. Human bones older than a million years old have been found. But according to the Bible, we human beings were created only 6,000 years ago, and almost everyone of us knows somebody who still believes that, even though science proves otherwise.

All the religions talk about creation. All religious texts say: "He created". What does the term creation mean? How does it work, biologically, chemically, physically? By what means does He do it? Nobody knows! Therefore, creation is a concept only useful for those with blind faith, or those who believe without any knowledge of what creation is or how it works. Because it does not give us any knowledge, we cannot accept creation as an answer to the topic of this chapter. We are searching for knowledge; we want to understand the origins of the universe and ourselves and how everything works. Believing is *not* equal to knowing!

So it seems we are stuck with evolution, the theory that is widely accepted by our scientists, teachers and intellectuals. It is still only a theory and not proven, but it is already treated as a scientific fact. But before we investigate the theory of evolution in more detail, I would like to eliminate one great misunderstanding about Darwin. He rejected the idea of evolution, as it is understood today, even in writing. He did not like the (now common) interpretation of his work as a theory – that every living being, from the beginnings of life on earth, evolved by blind chance. He used the word evolution very often instead of what he was actually talking about: mutation, which means that a species changes again and again in order to adjust to a changed environment, or to stay alive by the law of the survival of the fittest.

But a dog always remains a dog, a horse remains a horse, and a fish remains a fish! Darwin never accepted the idea that every living thing evolves on its own, changing from a single cell into a primitive animal, then into a fish, a reptile, an ape, and finally into human form. He saw clearly that such a theory would mean that each animal would have to run around for a few million years, for example, with an eye not fully developed yet and was of course not working; blind. Plants would have to reproduce before their seeds were fully developed. The list of such impossible ways of development is endless.

Lately, even scientists have had their doubts because of the DNA research. The way DNA is designed makes it impossible to cross breed between species that are too different. And when we do so with tricks, and cross breed two similar species, a horse and a donkey for example, then the offspring, the mule, cannot reproduce. I have my own question about DNA. DNA is by far the most complex design in nature. How could it have been, with such a complex design, already present in the earliest primitive life forms on earth when life was, at the beginning, very simple?

There is another important thought about this grand theory. If evolution, as seen today, were a fact, then it should still be going on right now. The earth should be full of animals and plants that are still developing, but such is not the case. All over the world, every creature is absolutely perfect for its location, environment, and for the function it has to fulfill. All too often we think or believe that some being is imperfect, only to learn later that we just did not understand why it is the way it is. Our scientists are still very far away from understanding all the details of how nature works; it is too sophisticated. But they believe that all these complicated, clever, and often sneaky systems in nature, like DNA, the brain and hormone, just to name a few, evolved out of a single cell billions of years ago, all by itself and entirely by blind chance. One man put this belief system into these words: "Believing in evolution is the same as believing that a wind, blowing through a junkyard, can produce a Boeing 747, ready to fly."

Now let us look at a few selected things in nature and how they could have evolved on their own. Of course, there are millions of such examples, but we can only look at a few.

First let us take a look at the ear, which is a very complicated and sophisticated system. It would require a Ph.D. in engineering, electronics, physics and biology to design such a thing. But with our limited knowledge it is not possible for us to create such a system. (Even our computers cannot compete.) Nature is much smarter. Our ear is so sensitive that at some frequencies the eardrum vibrates in an area of less than one billionth of a centimeter, which is less than the diameter of a hydrogen atom (the smallest). And many animals have an even more sensitive hearing system than we do.

Connected to the eardrum are tiny bones, the hammer, anvil, and stirrup. They transfer and amplify the drums vibration to the liquid behind the so-called oval window. (Liquid is the key to this system because it cannot be compressed and transfers pressure changes without delay. That is why hydraulics react instantly.)

These tiny bones, which take up less space than the head of a tack, are fully formed when we are born and have to last for our whole lives. They are completely disconnected from the circulatory system of our body. Therefore they cannot grow or heal at anytime of our life. This condition is required because of the function these bones have to perform. And now let us apply logical reasoning. We have to agree that such a set-up must be planned, designed, and tested in advance! It cannot evolve by itself by chance. Trial and error evolution cannot explain this phenomenon.

Now let us investigate the organ that is most precious for most of us – the eye. I will deal only with the eye, not with the fact that almost one-third of our brain is used to process, interpret, and evaluate what the eye sees. And I will look only at the lens.

For comparison's sake, a brief discussion of man-made camera lenses would be helpful. The cheapest lenses are single lenses, and they are not very sharp. The different refraction indexes of the diverse colors cause color auras at the outline of objects. In short, single lenses are lousy optics. We need at least four lenses with different refraction indexes, mounted serially, to compensate for this problem. The necessity of focusing at different distances is another reason why we need at least four lenses. Our best and most expensive lenses have up to eight lens elements, some of them made from very expensive exotic glasses. That is our technology.

Our eye does all of this and even more with just one single lens. It can focus near and far with extreme sharpness, which is impossible for one of our single lenses. According to the rules of optics, such a performance is not within the range of a single lens. But our eye does it anyway. How?

The German physicist Hermann von Helmholtz did a lot of research on the optical properties of the eye in the middle of the nineteenth century. He was the first to explain the optical principle. The incoming light passes first through the cornea, which is a fixed lens, then through a liquid – which has a certain refraction index – located between cornea and lens. Because of the shape-changing lens on one side, this lens-shaped liquid works like a second lens. Muscles around the lens are connected to the rim by so-called zonules. When stretched flat the lens focuses for infinity, and when relaxed it is curved for close-up focus. Helmholtz conceived of the lens as a deformable bag of fluid because he did not know better.

Today we have powerful microscopes and in creating them we discovered that the lens consists of thousands of ribbon-like fibers. They interlock and nest within one another very much like the layers of an onion. Each layer has a different refraction index. In this complicated and highly sophisticated way our eyes perform better than our best camera lenses. It is an optical marvel! From the viewpoint of engineering, such result cannot be achieved by any method of trial and error, as evolution is understood. Such a marvel has to be designed! Our level of knowledge and technology is still not advanced enough to perform such a design.

A very strange idea is haunting the evolution theory. It is the belief that our sea mammals, dolphins, whales, and all the others, were land animals who went "back" into the sea. How could that have happened? In order to survive, the very first animal who got this strange idea had no time to develop the breathing hole on top of their neck, which is essential for these animals; it is the only working breathing solution for them. So we have to imagine this procedure: The animal takes one last breath through the mouth, jumps into the water, and takes the next breath through the instantaneous breathing hole. But this animal does not want to be the first and last to do so, so it will have to reproduce. This means that this first animal had to change its method of giving birth from head-first to tail-first. Born head-first, a baby whale would drown during the birth procedure before it could reach the surface for its first breath. It is easy to see that something is very fishy about this idea.

Of course, the common belief that life started in the sea, and later on the land animals evolved out of the sea, involves the same kind of problems, only the other way around. It is even worse because now we talk about sea creatures that get their oxygen through gills, which do not work in air. If a fish wants to leave the water and move on land, it has many problems. It must change from gills to mouth breathing, and it has to develop lungs instantly. To change fins into legs right away would also be very helpful. It is still fishy!

Let me also mention the egg. The shell is made up of tiny, cone-shaped parts that wedge strongly together. These cone-shaped parts are slightly wider on the outside, and stick in the surface of the egg like a cone-shaped cork in the neck of a bottle. This design makes the egg surprisingly strong in resisting pressure from the outside, so that the feet of the breeding animal will not break it. But the shell can be broken very easily from the inside, and this is an essential fact of life or death for the hatching chick. Therefore, the very first egg in the history of the chicken must have been designed in this clever way. There was no room for trial and error. (Now we can see that a very clever chicken and an equally clever rooster must have been first, designed in order to produce the first perfect egg.)

Basilisk lizards, commonly found near rivers and streams in tropical regions of the Americas, are also called Jesus Christ lizards because of their unique ability to sprint across water. It looks very comical when they speed across the surface of the water. Fringe surrounding the toes of the lizard's hind feet enables the lizard to float on an air cavity each time one of its feet slaps the water. Before the air cavity has time to close, the lizard pulls its foot up and prepares for the next step. The lizards can do that because their hind legs are of a special design, and the brain is programmed to perform this feat. Common sense tells us that this cannot develop step by step until it works. It has to be designed complete and perfect for the very first time.

We may wonder how a bird can fly by just moving the wings up and down. They have a secret. Out of the shaft of their wing feathers we find rows of barbs. The barbs are united by rows of barbules, interlocked by means of tiny hooks. This ensures that the surface of the feather remains unbroken and provides resistance to the passage of air during the downswing of the wing. But because the barbules are hooked to the barbs out of center, they open completely during the upswing of the wing and let the air pass through. That is design!

If we look at any details in nature, fauna or flora, we find that everything is so cleverly designed and with so much sophistication that even today our scientists are still far away from fully understanding how things work and especially why they are the way they are. There is too much genius behind everything in nature and our science is not able to cope with it completely, at least not yet. We do not even know everything that exists. And worst of all, millions of plants and animals do things that are clearly preplanned for a future they cannot have any knowledge about.

So far I have produced only logical arguments. But there is also a pure scientific way to check the facts. Let me show you two very good mathematical examples, out of millions of this kind. It is about the possibility of life arising out of a primordial soup, the very beginning of life according to the theory of evolution as understood today.

Hemoglobin is one of the most important, essential proteins of life. It is the red matter in the blood. It carries oxygen from the lungs to the tissues and the waste compound carbon dioxide from the tissues to the lungs – not an easy job. Only twenty different amino acids are used to make up this protein. But a total of 574 amino acid molecules combine into this very complex protein. (Different numbers from each of the twenty amino acids are used.)

If we assume that all twenty amino acids were available in this famous primordial soup, what is very unlikely then is the probability of 574 molecules to join together into hemoglobin by chance – 1 in 10650 permutations.

This number is not astronomical because astronomy does not have such big numbers! Some examples:

1. From the start of life on earth, about 2.5 billion years ago, we had only $10^{17}$ seconds available.
2. Estimated number of stars in the known universe: $10^{22}$.
3. Estimated number of atoms in the known universe: $10^{80}$.

These numbers are not very much in favor of evolution. But the creation of hemoglobin is not the only problem. The very first form of life already required DNA in order to reproduce.

What are the chances for DNA to be "created" in such a way? The smallest DNA we know, the T-4 phage, a microscopic small creature, will have a specificity (probability) of $10^{78,000}$.

We must not overlook Huxley's monkeys. There is a recurrent line of thought among both lay people and non-mathematical scientists that, provided one had the right basic atoms and molecules at hand, then life could evolve by accidental aggregation and random mutations combined with the Darwinian principle of the survival of the fittest. To illustrate the point it has been suggested (by Sir Julian Huxley) that if millions of monkeys were to type in random fashion for millions of years, they would one day type a Shakespeare sonnet.

Almost everybody liked it, and believed it. But when some of the leading mathematicians saw it and checked the probability, they found that even in a million billion years it will not work. Or in one word: impossible!

All of this leaves us with only one problem: Is evolution by chance impossible or could it never happen?

Now, where are we? The creation stories of the many religions do not make much sense and do not satisfy our question of *how* and *why*. We can also prove beyond reasonable doubt that evolution by chance is impossible. Because of these facts, and because there are these two ideas only, I hear so often the question: "Is there anything else?" Yes, there is!

What I cannot understand, after reading and hearing so much during my life, is the fact that so far nobody has come up with the only possible and logical answer to the problem: *design*!

Design is the answer. Evolution cannot design; it needs a *mind*! It is exactly like the statement of Sir Arthur Eddington in the 1930 Cambridge Club: "The stuff of the world is mind-stuff." Matter is not a thing, but it is a *state of organization*, and the organizer is the mind of the causal cosmic intelligence.

Even though evolution is considered only a theory without proof, there are strong movements to teach it as fact and not as a theory anymore. And if you are in college and studying biology, do not dare to have another idea besides the evolution myth. I would really like to know, why our educational system is so narrow-minded. (It is the same with the Big Bang theory for students of astronomy.) Could it be that it is easier to cling to a scientific dogma than to be open-minded?

We can clearly see that everything in nature is planned and perfectly designed for purpose, location, and proper interplay with the environment. And each life form is a perfect design to match the geological time era and climate of our earth at any given time during the evolution of our planet. And this evolution was not always smooth. There were many disasters – ice ages, asteroid hits, pole shifts and overheating, just to name a few. Each time many species were lost, but right after every cataclysm life came always back with new forms, perfectly designed for the new conditions on earth. Therefore, none of the life forms of today had the whole 2.5 billion years of evolution available. They had only the time from the last catastrophe until today, 60 million years at the most. (Except some sea animals.)

How can anybody overlook these conditions? Every living thing is perfect in design from the very beginning and does not have to evolve. Living beings will always *mutate* in order to be more competitive, or to adjust to a changing environment, but a cat remains always a cat, and an ant will always be an ant. By the way, the ant is one of the few animals so perfectly designed, that they survive all climatic changes in the history of our earth. They never had to mutate during their time of existence; they remained unchanged. Their job within the framework of nature is one of the most important ones. Without ants our planet would be a biological junkyard.

Now we are hearing the final question: WHO DID IT? But before we find out let us take a break and have a short look at ourselves. What are we supposed to be, from a biological design viewpoint?

First our teeth. They do not resemble those of a carnivorous animal, the herbivorous nor the omnivorous. But they resemble exactly the teeth of frugivorous animals – fruit eater!

Another thing to observe is the length of the human bowel in the digestive system. Our anatomy books tell us that our bowels are three to five times the length of the body, which is measured from the top of the head to the soles. This is completely wrong for the purpose of comparison, because animals are measured from the mouth to the anus. If we apply the same rule to our bodies, then we have ten to twelve times the length of our body, which is the same ratio as in fruit eating animals. And the principle of digesting is also exactly the same.

Now we add the fact that most people cannot stand to see the flow of blood, and most of us are repelled by the smell of raw meat; we can only eat it if it is spiced and cooked. In contrast, the sight and smell of fresh and ripe fruit is mouth watering.

These observations should tell us that we are designed to be fruit eaters. Our taste and eating habits have been perverted so much through unnatural living over many millennia, that we should not be surprised at having so many diseases.

Now let us go back to the drawing board. Who designed all these living things? Who did it? These millions (or billions?) of species over the history of the earth are surely a lot of work, a big job. Some people may say that this is a design program beyond imagination. I do not think so. If we consider that we humans designed and produced trillions of different items of stuff during our history, then the number of different plants and animals on our earth appears not so high anymore. But it is still a big job, especially if we consider the complexity of each creature.

If we assume that every living thing on earth is inhabited, guided, or controlled by a spirit, then we should have this question: DNA is called the blueprint for the finished being or life form, but who reads that blueprint? A blueprint that is so tiny that only the most powerful microscopes can make it visible, but not reveal the billions of messages hidden within. Who can read and decipher billions of data and follow up on the encoded order? It is a marvel of biotechnology! Only a spiritual entity that consists of the finest "particles" of the cosmos, the units of God's thought in fact, is able to read the DNA easily in all its details. Therefore we must reason that a spirit must be the controlling force of development from sperm or pollen to the finished being. Edgar Cayce's work points in that direction; his spiritual self communicated with the soul of the people he helped. These souls knew everything about their body, and Edgar Cayce's soul got the information from them, and not from the body!

The movie *Jurassic Park* was a very good example of how much the properties of DNA can be misunderstood. Our scientists know already that from DNA alone we can never reconstruct an animal or in the case of the movie, a dinosaur. It requires a dinosaur mom, and a spirit that takes control of the egg for the development of the baby in the egg. But how does that work?

We postulated that an animal is occupied by a spirit, which then is in control of the body. Now we can extend this postulate: If a spirit can take over the control of an animal body and can read and reproduce the DNA in every new cell, then it seems to be very likely that this spirit is also in control over the whole growing process, and remains in control over the lifetime of the being. Dozens of reliable sources make this statement, including Edgar Cayce.

The next logical conclusion is this: If the occupying spirit controls a life form from the moment of fertilization or conception, then it seems to be very clear that both this certain spirit and the certain physical life form belong together. Why? Because these certain spirits designed their physical life forms in the first place! This is the only logical explanation that makes sense. Every life form on earth was designed and developed by that group of spiritual entities who wanted to occupy and control it. Of course, all of this design business had to be in full cooperation with all other spirit beings, and with the superior spirit-entity in charge of the planet, any planet.

It is easy to understand if we consider what we do. We design cars we want to occupy and control. Why? To have activities and experiences we cannot have with our bodies alone! Now we can go sixty miles per hour, or just enjoy the ride.

Exactly the same idea applies to the designing spiritual entities. To experience the physical world with the senses, what is not possible for the spirit form, they have created plant and animal life.

We can already see that animals and plants cannot design themselves or evolve from almost nothing into something; only the spiritual beings can do that. We humans can only play with what there already is, and do so pretty well sometimes, but we cannot design and build new species. We can breed mutations, but we cannot create.

When our "expert" scientists do not know why or how an animal does unexplainable things, they have a great explanation, or rather, excuse: *instinct*. This magic word eliminates for them the need to know what is going on. It is a very practical invention, this instinct. The previous paragraphs and chapters provided a foundation for a final explanation of what instinct is:

INSTINCT IS WHAT THE GUIDING SPIRIT KNOWS!

A very good proof of this statement about instinct is provided by the grunions. Grunions are small fish who spawn during high spring tides along the beaches of California and Mexico. They come in with the highest wave of the day, move farther up, make a hole in the sand and spawn into it. These holes are located exactly where the highest wave will reach after a few weeks, when the young fish hatch. The little fish hatch out of the sand in exact timing with the highest wave and are carried out into the ocean. If that wave is not as high as "was known" a few weeks ago, they will die, but this is very seldom the case.

One night in the early seventies, grunions put their eggs so extremely high that everybody believed they had miscalculated. No spring tide ever reached so far up. But when the time arrived, an exceptionally high wave came in and took the little fish out into the sea. This extra high wave was caused by an earthquake far out in the Pacific. Of course, the grunions did not know that; the guiding spirit knew! This is the most blatant example of instinct I know of.

Nature even has a very cute example for us. It is absolutely impossible for the autumn spider, *Metellina Segmentata*, to approach the female autumn spider without being eaten. As soon as he touches the net, the female rushes on and devours him. How do they mate? The male knows how to do it and survive. He does not touch the net. He waits outside for hours, often for days, until some bug gets stuck in her net. Within that short "window of opportunity", when she has her mouth full, he quickly approaches her, mates, and gets away quickly.

Such survival strategy is more than just reasoning. It also requires planning ahead and calculating the consequences. This implies high cognitive power that spiders do not have. Nobody can convince me that this is programmed in the spider's DNA or is a job of the tiny brain of this spider. He can never learn about the danger by experience. At the first attempt to learn, he is dead. No, this is much more than just instinct.

I think it has become clear by now where our bodies originated. We designed them for ourselves! "WE" does not mean that you designed yours and I did mine. "WE" stands for all human souls in the cosmos. And the job was done eons ago, long before our earth existed. We designed these bodies for the purpose of incarnating (transferring our consciousness), and experience the physical world. To smell the flowers, enjoy the rain, walk through a forest, and eat sweet fruits – experiences that we cannot have while only in one of our spiritual forms of existence. We did so as co-creators of the great causal intelligence of the cosmos. When we are in our spiritual form, we can do so as a part of this omnipresent and omnipotent force, which we are accustomed to calling God.

Because we are *human* souls, we designed our bodies with a brain that is capable of serving all capabilities of our souls. These capabilities have absolutely nothing to do with the idea of "survival of the fittest" *a la* Darwin. We can compose, play and enjoy music. We can build and comprehend a world of physics, mathematics, chemistry; the list is long. We are also on our way to understanding and comprehending the universe. We can write and read books, even fiction. This list of examples could go on for many pages. But all of these capacities have one thing in common. They are not required for survival of the fittest! And many of them were dormant for millennia. We did not develop these capacities in our time; we only learned how to use something that was always there.

If we are not just a physical body, but in reality a spiritual entity, a soul, then the question of our age is of great interest. But before we can talk about a thing as age, we have to know and define what time is.

Time is something as strange as space. Both of them are beyond comprehension. We do not know what they really are. So first let us take a look at space.

Logical reasoning tells us that the cosmos must be infinite. (The exotic ideas of some scientists do not count if the topic is logic and common sense.) When we assume space to be finite, we face this problem: If there is an end to the size of the cosmos, then what is behind the end? What lies outside the cosmos? Albert Einstein's theories hold that the universe is curved and therefore a finite entity. He did not care about the question of what is outside that finite universe. Therefore I do not buy this idea of Einstein's; it is against common sense.

The only possible solution out of this dilemma that I can see is to postulate that infinite space has an infinite number of universes in an infinite cosmos. Of course, this is impossible to conceive. And as a bonus, we have another inconceivable fact: Wherever you are in an infinite space, you are always right in the middle. Einstein, who also was confronted with the possibility of infinity, had a very nice and clever way of helping a non-scientist get this idea. He compared an infinite space with a two-dimensional infinity, the surface of a globe. At every spot on this globe you can consider yourself as being in the center. Of course, this is a mental crutch, but a nice one, and easy to comprehend.

And now time. Time is as odd as space and cannot be comprehended either. Again, by logic, time also must be infinite, without a beginning or an end. The same problem with space confronts us with time. If we think of time as being finite, having a beginning and an end, then what happened before the beginning and what will happen after the end? It is exactly the same dilemma. Another strange property of time is the fact that there seems to be no such thing as the present. One second ago is still past, and one second from *now* is still in the future. We can narrow down this thought game to milliseconds, microseconds, nanoseconds, and even an imaginary short time; we will always only have a past and a future. A real present, a "now", does not exist because time is a moving entity and cannot be fixed into a "now" condition.

As a result of all this reasoning we now have a nice collection of three possibilities. Take your pick, each one could be right, each one is written somewhere in a book.

First, *now* is not possible. Second, any given moment in time, even the distant past or future, is always *now*. Third, time does not exist because something that does not have a present cannot be. If we think about these three statements, then we may understand the many sources who tell us that time is an illusion. The number of "beautiful" mathematical theorems and formulas that a physicist can build around these paradoxes is practically unlimited. (Einstein was one of the many who ran into this trap.)

Einstein's famous space-time continuum (three dimensions of space and that of time a four-dimensional continuum) means that time serves only as a factor to measure the relation and interaction of bodies in space. Take the physical universe away and time ceases to exist! I wonder if Einstein knew about Edgar Cayce. If he did, then he must have been very surprised about a fantastic similarity. Cayce stated in one of his readings that space and time are an inescapable basic requirement for the existence of a physical world. He said that God first provided space and time in order to make the creation of a physical universe possible.

Now we have the opportunity to apply our tool of logical reasoning. If we think backwards about this reading by Edgar Cayce, we will arrive at an interesting conclusion. If God had to create time before He could start making a cosmos then we have to assume that before then, there was no time, which means that God is timeless by nature. He is eternal. And if we are a part of Him, a cell in His body, then we have to be subject to the same conditions.

This is the answer to the question of how old we are: Because we are part of God, an individualized spirit, a living soul, we must be of the same "age" as Him – ETERNAL! (This, of course, means that the religious promises of salvation, of being saved, is not based on facts and only an invention of the church in order to keep us bonded.) Later on in this book we will see the consequences and responsibilities which result from this relationship to God and the universe. We will see that we have to progress towards perfection on our own; nobody can do it for us. Not even Jesus, as so many people believe.

Existing for eternity is not boring! The next chapter will show that living an eternity is not dull for us. Just the opposite is true; we are very busy and happy during all this time.

# Chapter V
# AN ETERNAL JOB

> The meaning of life is that it makes no sense to say that life has no meaning.
>
> Niels Bohr

We are not goofing off here on earth; at least most of us are not, so why should we goof off for eternity? Many people are looking forward to goofing off in heaven. Some (monks) spend their whole life in preparation for that idle time. They will be very much disappointed. First of all, there is no such thing as the heaven promised by most religions and, of course, a hell is also just a man-made invention. We will talk about that later. After the death of the body, these people will be surprised that there are no pearly gates, no wings for us to fly around, and no clouds to sit on and play harp for eternity. This would be a boring existence. No! All of us are busy with useful work to do, and we love it.

But first we need to know a few more details about what we are. We consist of three parts when incarnated in a physical body. After leaving that body (discarnated) we are made up of two, and when we are in our original form back to the side of the causal cosmic intelligence, or God, only one.

While here on earth, we are made up of the physical, astral and ethereal body. The physical body has the physical mind, or just mind, which is the intellectual ability of our brain. The astral level has the subconscious mind, and the ethereal body is home to the superconscious mind. Our real self, the ethereal entity, is in direct contact with the great causal intelligence of the cosmos, or God. It is pure spirit, nothing else. It is a soul. All souls are interconnected by a great recording and information system that we cannot comprehend, known as the so-called *Akashic Records*. Edgar Cayce called them sometimes also "the book of Life", and Paramahansa Yogananda wrote about them in his books. (The colloquial usage of Akash in the Hindi language represents the sky. As per Hinduism it is a supposed all-pervading field in the ether in which a record of past events is imprinted.)

Each human entity, or soul, can dig into these records and look up any part, but this is only possible from the level of the etheric. If the entity is incarnated in a physical body, the mind of the brain cannot normally do it, but some people like Edgar Cayce had the capability to reach the superconscious level in trance and tap this infinite source of knowledge.

The ethereal level, or etheric, is the highest level that each one of us can and will reach, even if it takes a long time to get there for some of us. Actually, this highest level is where we came from; we were created ethereal beings. We are just creeping back right now!

Our second body from top is the astral body. It holds the subconscious mind. When we are incarnated into the astral body, our consciousness moves from the ethereal down into this body, and we identify ourselves with it. The astral is also a kind of physical world, very similar to ours, but the matter is so fine, and the vibration so high, that we cannot detect it from our gross physical world. Astral is the plane where most of us go when we "die" here on earth. Then our consciousness moves from the physical to the astral. We do not die; we make a transition!

Our third part is the physical body with its mind. What this mind does and can do is, to a high degree, a function of the brain, but there are also strong connections to the subconscious and the superconscious minds.

Because we are incarnated and our consciousness is located in our physical body, most people are fooled and believe that this is all there is; they identify themselves with this body. We already know that this is an illusion.

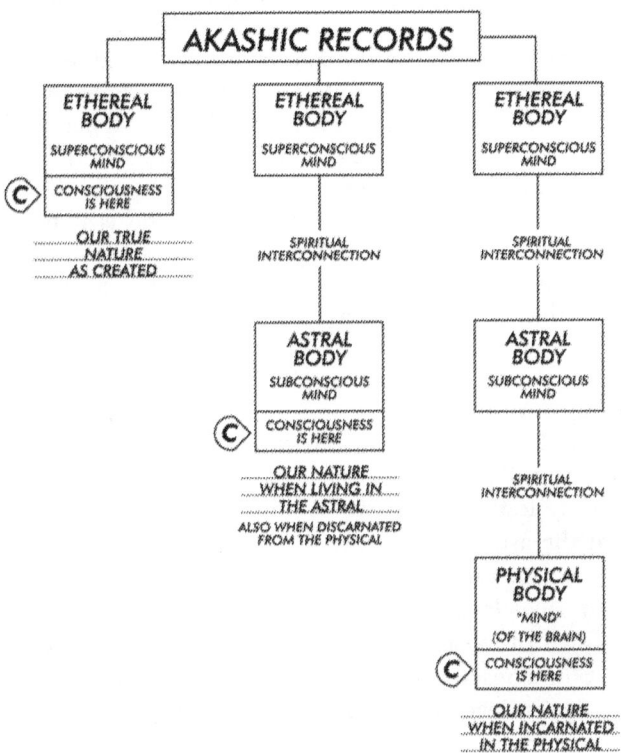

FIGURE I: OUR THREEFOLD NATURE

Our problem is the fact that we are only one entity and therefore have only one consciousness, even though in the worst case, when we are here on earth, we consist of three mind levels. We automatically identify ourselves always with the body we are born with, the place where our consciousness is located. When we live in the etheric, as ethereal beings, we feel that this is all there is of us, because our consciousness is there. When living in the astral realm, we are conscious of this and believe that this is all there is, but it is not. There is still the superconsciousness of the ethereal body above us. And when incarnated here in the physical, we think that this is the real thing. It is not! There is still the subconsciousness (astral) and the superconsciousness (ethereal) above the location of our consciousness. More and more people are finding out about their true selves at the end of the last century, and learning how to connect to their own higher selves in meditation and prayer.

From the lower realms we cannot detect the higher ones, but from the astral we can detect the physical, and from the ethereal level we can look all the way down to the other two.

Our normal level is the ethereal, the soul level. It is the highest form of existences; it is the form in which we were created. Our consciousness is in our soul and that is it. No other body or level of existence is involved. We are pure spiritual energy in this stage. (Remember, God created man in His image, which is not a physical human form.) When in the etheric – and we are there most of the time under normal conditions – we are very busy with the task we are assigned to do. We do our work in full cooperation with all others and accept the guidance of our boss, the universal causal mind and His executives. We do this of our own free will, and we love to do it; everything runs in perfect harmony. To be sure, I talk here about us in the pure soul state of the etheric, and not about anyone in the astral or physical world, and especially not about those entities who completely failed on the earth (mass murders and war criminals for example). These entities receive a special treatment right after their transition into the astral. As I have already mentioned in Chapter I, when I talked about

NDEs, these poor souls are subject to a shock treatment in the form of a hell-like experience. And now, what are we doing?

Our lives focus on the development, design, creation and maintenance of everything in the physical and astral realms of the whole cosmos, from bacteria to animals, to planets, to solar systems, to galaxies – everything! We work as co-creators of the universal mind, God. We help bring His ideas and desires into shape. Some of us are busy with the planning and development of new planets, and get them ready to harbor life. Other groups determine what is needed and then design and create the required life forms. Of course, all of this work is done by teams, which often count by the millions. Most of the time, to be sure, the creation of new life forms will not be required. Established and proven designs from other, older planets will just be transferred to a new planet. Many proven and tested design principles will be repeated over and over. Just like in our own technology, certain things cannot be made better; four legs, joint designs, the principle of the eyes, ears, breathing holes for sea mammals, skeleton mechanics – the list is endless. And because time is infinite, the whole thing is a never-ending job.

The smarter entities among us take care of suns and other big items. There is a hierarchy of the universe, just as here on earth. The most advanced among us could be called the executives of God's cosmos. They do the planning and organizing and handle all required social functions. They are the leaders of planets, solar systems, star groups, galaxies, and galaxy clusters. And there is one for the whole universe, one who is not God. Later, in Chapter IX we will see why each universe has one. (If I understand the various sources right, then the Christ-entity is in charge of our planet for all people of the whole history of the earth. Each planet has one.)

I am sure that this picture of an organized and maintained universe makes a lot more sense than the naive question about other life in the universe besides ours. "Are we alone?" is in my opinion the ultimate ignorant question. It is the worst of all dumb questions possible. Why? Because it requires the belief that the billions of galaxies with a total of quintillions of suns may exist only for our entertainment, only serving as conversation pieces for us. And it entertains the thought that our tiny dust particle, the earth, is the only thing of importance in the universe. It is a ridiculous thought!

I have one word for the astronomers and cosmologists. With our modern technology we will observe many events that defy explanation through known physical laws. But if we consider the possibility that manipulating by ethereal powers is behind such cases, then we may understand.

Because love is the strongest mental force or attribute of the etheric, everything runs smoothly, in peace and total harmony, and is fully enjoyed by everyone involved.

There are many highly developed physical super-civilizations in every galaxy where the people, in respect of their mental condition, are practically ethereal. Every person in these societies is permanently fully aware of the God-heritage. And even though they are in physical bodies, they command many of the powers of the ethereal realm. (According to Yogananda's books, some of them incarnate from time to time on earth in order to help us. The best example being Babaji, who has lived for many centuries in the Himalayas. In the 1930s he even had two Americans among his disciples. When he wants to or has to move to another location, he just says, "*Dera danda uthao*". (Let us lift our camp and staff). At that very instant, Babaji and all his disciples disappear from the old location and manifest at another location. This is called *astral travel*, and it is *not* fiction!

Generally, we have to believe that there is physical life like ours, or similar to ours, on every planet that could support it and, according to many different and independent sources, astral life exists (undetectable to us) on most of the others. Here I must mention another important fact. Two physical bodies cannot occupy the same space; we know that. However, astral bodies can occupy the same space that is already taken by a physical body. This means that an astral planet could occupy the same spot that our earth already fills. The same applies to suns and galaxies. And, of course, to us. Our astral bodies are always within and around our physical bodies wherever we go, except if we introduce an OBE. Now we can understand why sources like Edgar Cayce, Sri Yukteswar, and others state that all planets of our solar system are inhabited. They talk about astral life.

After all of this, I am sure you will have an important and logical question: If everything in the spiritual realms of the cosmos is love, peace and harmony, then what are we doing here on this unsafe planet that is anything but peaceful? Something must be wrong. Maybe with us?

In our original, ethereal-only state; our soul is an individualized spirit with a free will and tremendous powers. As long as we use these powers as cooperative co-creators, everything is fine and dandy. But some entities are not capable of fully accepting the laws and rules of the cosmos. They take advantage of their free will and begin to misuse their powers. They separate their mind from the mind of God, and start to do their own thing. Some even become so ignorant as to believe that they are as powerful as God and try to play God, often working directly against Him and all rules He has set.

Such behavior is against cosmic law and cannot be tolerated! When we separate ourselves from the cosmic unity, this law goes into effect. Of course, God does not have a death penalty for any soul. How could He destroy a part of Himself? Such bad behavior is considered only to be a flaw of character, which can be corrected. And He has provided a sure way to do just that. It is a procedure of retraining which makes sure that the development to the normal state will be genuine in the mind of that soul and not calculated, based on the knowledge of what is going on. This requires amnesia of the soul memory! (The spiritual memory.)

These "schools" are pretty rough, but they work. Our earth is one of them! Each galaxy has as many of such planets as required. These "bad" souls we just talked about have to incarnate on such a planet with all knowledge of their true identity eliminated (the memory blocked/amnesia), and they will be subject to the law of karma, the spiritual law of cause and effect. Among the many sources for this knowledge, Edgar Cayce provided the most details about this law, and Ian Stevenson, M.D. found and investigated the most cases that provided evidence. And, of course, the Eastern religions treat reincarnation as a fact.

Let me repeat: This kind of amnesia is required for a genuine improvement of character. Otherwise we could calculate everything we do because we know what is going on. This also means that reading this paragraph can have two results. You are not ready for this and therefore cannot accept it, or you sense with your subconscious mind that you are reading the truth, and you realize that you must be almost "through" with your school time here on earth because you now understand why everything is the way it is.

These "school planets" provide everything that is required for the proper development of the souls who go astray. They provide the greatest beauty as well as ugliness and cruelty and everything in between. Different ethnicities are created to further challenge these troubled souls. We are made to learn the hard way! Most of these "schools" are on planets that still have very unstable geologies, causing natural disasters again and again. They are too young and not stable enough for highly developed super-civilizations, but good enough and just right for the purpose of correcting souls.

These lost souls have to incarnate and reincarnate on these planets until they first suspect, then learn, and finally know what is going on. (Example: Reading this and other such books.) Knowledge is the goal, not just faith or believing in something. The most important factor is the law of karma (Sanskrit for deed or fate), the law of cause and effect. It is a very simple, mandatory law. Whatever bad things you do in your life will hit you back in a similar way if not in this life;

then in one of the next ones. Whatever good you do, will be rewarded likewise. The law of force and counterforce in physics is actually the equivalent in the physical world.

Finally, when we have worked out all of our karma and apply the cosmic law of love in everyday life, we do not have to reincarnate anymore and we can go back to our true heritage.

Unwillingly I have just given away the reason for the existence of reincarnation on our planet. There will be an extra chapter on this important topic. I also realize that this whole book spoils the nice fairy tale belief systems of our religions. The cosmos is real, not cute! And I also think that it is clear by now that all of us belong to the "lost souls" who tried to do without God. If not, we would not be here. (Exceptions will be explained later.)

One thing should be clear as a result of logical reasoning: We better forget about a heaven on earth in the near future. This is not the purpose of this planet, not yet. There is a steady stream of "newcomers", who are still extremely ignorant and spoil the life of the already advanced people who will be able to leave this planet soon. This planet runs by the rule that "in history we can learn that mankind never learns from history", as my history teacher used to say. Of course, it is our task here on earth to always try to do our best. We should never give up, because we have to concentrate on nothing but our own development by always caring about and helping other people. This sounds like a paradox, but that is the way it is because such behavior requires *love*!

So as not to discourage us completely, we get a break from time to time – a few decades or even centuries of peace, for example. Of course, such a period of peace will never be global, only in some areas. This system assures that all the bad newcomers will have a place to start their journey on earth where they can best correct their character flaws and work out their karma. And it also provides better places for the entities who are already on their way back to normal to a great extent, the ones that are almost corrected. Every entity incarnates in an area and under conditions that is best for their personal karma and requirements for learning.

We must know that every condition, and all the circumstances we find here on earth, were planned *before* we incarnated and it is the way it is for a reason – karma! For example, if we live in good health and without financial worries, then this could be a reward for doing good in a past life, but we better watch out! This will also be a test for us. Will we care about other, less fortunate people and help them because we can, or will we just live for our own advancement and satisfaction? Just see what some of our very rich people *do not do* with their money. They can be sure that their next life will be under opposite conditions!

Now we know what Niels Bohr was saying when he made the statement that life does have a meaning. And we also know that eternal life is not boring at all. But most important is our knowledge that this life on earth is not a one-shot deal without meaning. It is only a very small, but important, episode of our existence.

Millions of people reject the existence of God by the following reasoning: If there is a God, a God who loves every human being, then why does He permit all the poverty, misery, mishap, cruelty, killings, wars, and natural disasters? All of this speaks against the idea of a living, loving God. But when we apply the law of karma and it is consequences, then suddenly everything makes sense and we understand that pure love is behind all of this, even if it is hard to imagine.

So let us end this chapter with a sentence from the Edgar Cayce reading 3245-1:

> *Know* that you are going through a period of testing. Remain true to all that has been committed to thee, and *know* that each day is an opportunity, and an experience.

Please observe that twice he said know. There is not a single request for faith or belief. Knowledge is the goal!

One more thing. You may think that the content of this

chapter is so strange and weird that you cannot believe it. In order to prepare you for the rest of this book, I would like to provide a short list of *facts*, each one fully documented in many different ways:

1. Hundreds of so-called crop circles develop in crop fields all over the world each year. They always appear overnight, and the geometric designs are so complex and beautiful that nobody can fake them in the dark. They are fact, but we do not know who does it, how it is done, and why.
2. Instantaneous human combustion also happens all over the world. Photographs show that the body of the person changed into a small pile of dark ash. Very often an arm or a leg still remains at the side. Cremation at the highest temperatures over many days cannot do that. The surrounding parts of the house, however, are never ignited. What is it? We just do not know!
3. There were many monks and nuns who did not walk the steps of the stairs of their monasteries, they floated up or down. They were told by the Catholic Church to stop that and walk like all the others do, so they did. But how did they do that?

These three examples should serve the purpose; I know many more. Because these are proven facts, we can make the following statement: We are still very far away from understanding the universe and its possibilities! It is weirder than we can think!

# Chapter VI
# REINCARNATION: FACT OR MYTH?

> Absence of evidence is not evidence of absence.
>
> Old proverb

Benjamin Franklin, when he was twenty-three years old, wrote his own epitaph which, maybe for its beauty, became a kind of classic.

*The body of*
*Benjamin Franklin*
*Printer*
*Like the cover of an old book*
*Its contents worn out,*
*And stripped of its lettering and gilding,*
*Lies here, food for worms.*
*But the work shall not be lost,*
*For it will, as he believed, appear once more,*
*In a new and more elegant edition,*
*Revised and corrected*
*By the Author.*

Of course, this is no proof for reincarnation in any way. But I think it is interesting that this man, who was a kind of Jack-of-all-trades genius believed that he would be back later on, an improved version of himself. And so did many other great men: Giodarno Bruno, Caesar, R. W. Emerson, Goethe, Lessing, Nietzsche, Plato, Pythagoras, Schopenhauer, Shakespeare, Voltaire, Oscar Wilde, Jack London and Henry Ford, just to name a few.

Because this topic can only be handled with the tool of logical reasoning, most people who have investigated this field had one common problem: they had a lot to lose!

Most of these investigators were members of certain academic groups, like doctors, doctorate-holders, clerics and others. In order not to be ridiculed by their colleagues they could not afford to state any particular case as definite proof of reincarnation. For most of them, it would be the end of their career. So they always concluded at the end of their case histories, that the case was *suggestive* of reincarnation, even when there was no other explanation possible. Very few made clear, positive statements. Some of them were really great minds like Immanuel Kant, Arthur Schopenhauer, Leonardo da Vinci, and many others. Leonardo even stated openly that he was permanently haunted by memories of past incarnations.

First let us find out what speaks against reincarnation. All we can find is opinion and supposed lack of evidence. "It is not in the Bible, and so it cannot be!" is what we hear most of the time. We will see that this is not true. I could not find one single logical argument against it, only belief and opinion.

But we have hundreds of well-documented cases that speak in favor of reincarnation. So let us take a look at a few of them that are officially verified.

A Liverpool (U.K.) physician hypnotized a housewife, Annie Baker. She had never been to France, and she did not understand any French, nor had she any interest in the language. But when under hypnosis, she spoke a perfect French. She talked about Marie Antoinette as if it just had happened, and she gave her own name as Marielle Pacasse and her husband's as Jules. She also gave the name of the street in which she lived as Rue de St. Pierre, and said it was near the Notre Dame Cathedral in Paris.

Investigators went to Paris, but they could not find such a street. But further search through the city records verified that at the time in question there had been a Rue de St. Pierre in that vicinity.

Also under hypnosis, she did not understand what tea was.

As a matter of fact, tea was almost unknown in France of 1794. I wonder if anyone has another explanation for this besides reincarnation.

There are hundreds of others like this one, but the star of all cases is the girl Shanti Devi. She was born in Delhi, India, in 1926. At the age of three, she began to talk about a life she had at the town of Muttra, about eighty miles away. She said that she had been born in 1902, her name was Lugdi, and that she was a Brahmin by caste. She stated that she was married to a cloth merchant named Kedar Nath Chaubey. She also said that she gave birth to a son and died ten days later.

Her parents tried to stop these tales, but instead it got worse and worse. When she was nine, her family wrote a letter to Muttra to find out if such a person really existed. Kedar answered the letter and confirmed all of Shanti's statements. Then he sent a relative to the girl's home, and Shanti identified this person immediately. A few days later he himself came unannounced and Shanti identified him as her "husband" on the spot. Now, her parents realized that something real seemed to be behind the tales of their daughter, and not just wild fantasy and imagination. They contacted the government. First, the government officials established that Shanti had never left Delhi, then they appointed a committee to escort her on a trip to Muttra and note everything down for the official records. (We must understand that the government of India is very open-minded in such cases.)

When Shanti left the railroad station in Muttra, she recognized a past relative in the large crowd. Then she was seated in a carriage and the driver received order to follow her instructions. She first led them to Kedar's district and house, even though it had been repainted in a different color. An old Brahmin appeared near the house, and Shanti identified him correctly as her previous father-in-law, Kedar's father. Then the whole party entered the house. Here Shanti answered many questions regarding the arrangement of the rooms, closets, built-ins, and other things. Each one of her answers was absolutely correct.

Then they went to the house of her "previous" parents, whom she correctly identified and called by name out of a crowd of more than fifty people. And again, she answered all the questions about the house correctly.

From there she led the group to another house, which was the home of Kedar Nath Chaubey's family. There she pointed to a corner in one of the rooms and said that she had hidden money there. They lifted the floorboards and found a vessel to keep valuables, but it was empty. Shanti insisted that she had hidden some money in this secret place. At this point, Kedar acknowledged that he had found and removed the money after his wife's death.

Shanti also used idioms of speech typical only for Muttra. Her use of this dialect was especially interesting for the government officials. For them this was sure proof of Shanti's case of reincarnation. Shanti had the best accuracy record among the many cases in the world. Every single statement she made could be verified and none was incorrect!

I would like to hear about any other explanation for this case besides reincarnation. It is much more than just "suggestive". And I would like to use some logical reasoning right here. The girl Shanti was born with her own brain, and not with the brain of Lugdi Chaubey which collected all the memories of her life in Muttra. This means that there must be a third memory system besides the brains of Lugdi and Shanti! It is the memory of the soul!

I think that these two case histories should be enough for our purpose. If you are interested to learn more about such cases, there is plenty of literature about it in our bookstores. (See *Suggested Reading*)

We know that some Eastern religions, Buddhism and others, not only accept reincarnation, but teach it as a fact. But what about the Bible? Almost everyone tries to tell me that the Bible says nothing about reincarnation. In a way they are right, because there should be nothing about reincarnation in this book.

Why? The early Christian church accepted reincarnation as a fact, but then there was the Council of Constantinople in A.D. 553, over 500 years after the life of Jesus. There, by a tight vote of three to two, it was decided that everything about reincarnation must be discarded and removed from the Bible. Why did they do it? For what reason?

The first centuries of the Christian church showed clearly that knowledge of reincarnation was detrimental to the clergy's control of their followers. They could not be scared by the fictitious penalties of a hell, and they could not be lured into blindly following by the promise of being "saved" and of salvation in an imaginary heaven. The people knew that heaven and hell, as pictured by the church, did not exist. They also knew that a paid middleman between them and God was not required. They observed direct contact with God by doing their own prayers in their home, just as Jesus had advised. Therefore, reincarnation had to be eliminated from the Bible, the basis of the teachings of the church, so the church could survive as an institution.

But whoever did the job of "cleaning out" the Bible was not thorough enough. Many passages on reincarnation slipped through and remained in the Bible. A few good examples:

> And they asked him, saying, why say the scribes that Elias must first come? And he answered and told them, Elias verily cometh first, and restoreth all things; and how it is written of the Son of man, that he must suffer many things, and be set at nought? But I say unto you, That Elias is indeed come, and they have done unto him whatsoever they listed, as it is written of him.

Even if you read the above quotation from Mark 9:11–13 ten times, it is still not easy to understand, because the King James translation of the Bible does not make for easy reading. Scholars of this Bible say it is clear for them that he talked about John the Baptist. But there is help for us. It is the Lamsa Bible which I already mentioned in Chapter III. Lamsa translated directly from the Aramaic originals, and not from other translations. His Matt. 17:12–13 reads:

> But I say to you, Elijah has already come and they did not know him, and they did to him whatever they pleased. Thus also the son of man is bound to suffer from them. Then the disciples understood that what he had told them was about John the Baptist.

Each of the two Bibles says that John the Baptist was a reincarnation of Elijah.

Also of note is John 9:2 in which a man who was *born blind* was brought to Jesus and his disciples asked him, "Master, who did sin, this man or his parents, that he was born blind?"

We must not dwell on the answer, but rather consider the wording of this question. First, his disciples asked this question. They were the students of Jesus; therefore this question had to be in agreement with his teachings. And he did not correct them! Second, if they assumed that a man born blind could have been punished by blindness for sins he committed, then these sins must have been committed in a previous life. And the fact that they also mentioned the possibility of sins by the parents shows that they were very well informed about the law of karma, by which parents can be tested by being required to love a crippled child.

In considering the Bible there are many unexpected references to reincarnation that can be found by just applying reasoning and common knowledge. Some references are only indirect, but nevertheless of importance for a complete picture.

Every Christian has the question: Where was Jesus after age twelve until he was thirty? The Bible is silent about this period in the life of Jesus. But with the help of a little general knowledge and some logical reasoning, we can find out.

The answer to this problem is hidden in Matt. 27:46! You may say: Wait a minute; those are his last words on the cross. That has nothing to do with the unaccounted for time between ages twelve to thirty. Let us see. What did he say? *Eli, Eli, Lama Sabachthani*. Some Bibles print that capitalized, some in italic type. Why? Because nobody knew what that meant. Neither the writers nor the translators of the Bible understood these words. They are not Aramaic, a branch of Hebrew that Jesus spoke, nor are they any other language that was spoken in the whole Middle East or Mediterranean area in the time of Jesus,

So the Gospel writer made a guess as to what it could mean, and he was dead wrong. So, what is it? Colonel James Churchward and Don Antonio Batres Jaurequi (both prominent Maya scholars) found out, independent of each other. It was *Mayan*! Properly spelled it would read, *Hele, Hele, Lamat Zabac Ta Ni*. I think it is a miracle that the writer came so close after all these years. Well, what did Jesus say and why in Mayan? He said, "*Hele*: I faint, *Hele*: I faint, *Lamat Zabac Ta Ni*: Darkness is coming over my face."

We must admit that this makes a lot more sense. Jesus could never say, "My God, My God, why hast thou forsaken me?" as it is written in the Bible. Such words would be contrary to his teachings; God does not do that!

And now why Mayan? Who speaks Mayan? According to the books of James Churchward, the Mayas in Central America, and the Naga Mayas in the foothills of the Himalayas speak the Mayan language, and about half of the Japanese language is of Mayan roots. When visiting an old monastery in the Kashmir area of the Himalayas, an old rishi showed to him 2000-year-old temple records and informed him that Jesus was there for more then ten years, and he learned everything required for his mission. He learned the cosmic laws, who he really was, the best meditation techniques, and of course, Mayan.

Now we know where he was between the ages of twelve and thirty, why nobody in the Middle East could have known about it, and why Mayan words entered his mind at the moment he fainted. It was because he learned everything about his true self in Mayan. And we also know, just by logical reasoning, that in fact he did teach the wisdom of the Eastern religions, the very ones that so many intolerant Christians reject. (Remember, Buddha was born in 562 B.C. It was a Buddhist monastery.)

The records in this monastery also state that Jesus became the greatest master who ever spent time in this place. He mastered mind-over-matter tasks, which no one had done before or managed to do after him. From this it follows that the so-called miracles that Jesus performed, as written in the Gospels, were not miracles at all. The thought-energy of the causal cosmic intelligence (God) is the strongest force of the cosmos; it is the force that causes all things to happen. Jesus just knew how to use his portion of this force.

There is another fact about Jesus that is not in the Bible, but has very much to do with the "Maya story" just relayed. It is well-known in Europe but almost unknown here in America. Many books about this topic exist there. The German version *Starb Jesus In Kaschmir?* by Siegfried Obermeier is already in it is third print run. The English version *Jesus Died In Kashmir* was written in 1977 and published by Faber-Kaiser (London). Erich von Daniken wrote *Reise Nach Kiribati* in 1981, and Ernest B. Docker wrote *If Jesus Did Not Die On The Cross?* in 1920. There are many more. As you can see by the titles, these books are suppressed by the church with all available means. What is so very interesting is that these books make sense and are backed by numerous photographs.

Therefore I believe that you may be interested, too. It is a long and very detailed story; that is why we have whole books about it. I will condense the story down to the bare essentials. So, in a nutshell, this is the content of all these books:

When a man was executed by crucifixion at the time of Jesus, he died a long, painful death over a span of three to five days. Nobody died within one or two days. Jesus and the two other men had only hours to die, because at sunset a holy day began and they had to be removed by then. That is why they broke the legs of the other two men, so that they would die quickly (the "standard" method). Jesus, being an absolute master over matter, faked his death and so fooled the Roman soldiers. (Note that even today some Hindu ascetics in India put big hooks into their flesh and pull heavy loads with them. When the hooks are removed, there are no wounds and no blood is to be seen! If they can do it, Jesus could do even better.)

Joseph bribed Pontius Pilate to get permit to remove the body of Jesus. Joseph knew what was going on, and so laid him formally to rest. After three days Jesus was well enough to move. All his followers were then informed that they would find him in Galilee (Mark 16:7). Why Galilee?

1. Galilee was out of the sphere of influence of Pontius Pilate. If a living Jesus had shown up in Judea or Samaria, it would have been a political disaster.
2. Most of his followers, the best of them, were in Galilee.
3. It was on the route he had planned for his final mission.

Luke 24:39–43 removes all doubts that he was still alive as a physical being, and not a spiritual form. 24:39 reads, "Behold my hands and my feet, that it is I, myself: handle me, and see; for a spirit hath not flesh and bones, as ye see me have." 24:40–43 mentions Him asking for food and partaking of it to prove His state.

He then instructed his disciples about their missions and explained that he still had to take care of a lost tribe of Israel far away. (He really was the messiah!) Then he went to Kashmir.

In Kashmir, in the area of the city of Srinagar, most of the names of people and villages are definitely Jewish. If you see the photographs of these people – especially the children – you would believe that they were taken in Israel. Jesus was there between the ages of twelve and thirty to study in one of the many monasteries there, but he now returned as a teacher.

A very fancy shrine contains the tomb of Jesus; the books have many photographs of this building, inside and out. The people there, even though they are now Muslim, take very good care of the building and the tomb. Villagers know who is in this tomb, and they do not mind taking care of it because according to Islam, Jesus was a great prophet.

Erich von Daniken, author of *Gods from Outer Space*, was also in Kashmir and wrote the book *Trip to Kiribati* about what he saw. He believes that Jesus did not have to walk to Kashmir, that he got a ride from a flying machine (UFO?). He points to Acts 1:9–11. He did not go up, he was taken up. And who were the two men in white apparel, who knew so well what was going on?

The books talk about another grave. Moses, who led his people into a paradise-like, green valley (the promised land), and not into the Sinai Desert, was buried on the top of the Nebu Mountain in Kashmir. And like the tomb of Jesus, this grave is well kept today by Wali Reshi. His family has done this work for over 900 years. The people there know that in this grave rests the man who brought their ancestors into this land thousands of years ago.

What did Edgar Cayce say about reincarnation? More than all the other sources combined. For hundreds of people he gave so-called life readings, and indicated and detailed former incarnations of them. Most other sources just state that a certain incarnation occurred, or try to prove that incarnation is real. Edgar Cayce always gave much more. He explained the reason for each incarnation, and what they had to do with the present one. He also explained the relationship between the many incarnations of someone's past, and often also why they live together and interact with people in this life. All of these life readings were given under the viewpoint of karma, the spiritual law of cause and effect. Most of the time he told others whether a certain incarnation was positive (a gain), or negative (a loss) for the development of their soul. He also explained how the present life was a karmic follow-up to an earlier life. Then he gave advice about what the person should do about it. He did not just serve the curiosity of the people who requested the reading; his readings did help and teach. And he always had a definite aura of authority.

Naturally, the people around him wanted to know about the source of all this knowledge and wisdom. So, he held special readings to find out. The result: His consciousness left the physical body, passed through the different levels of the astral worlds, and ended up in the ethereal realm. There he tapped the *Akashic Records*, which contain *everything* ever thought and done by all of mankind, and also everything that ever happened to the earth and on the earth. In this state he could also find the bodies of other people, and communicate with the occupying soul (who knows everything about the body). From that soul he got a complete diagnosis much better and more thoroughly than any doctor could do. After that he would tap the infinite knowledge of the *Akashic Records* to find a remedy for the problem at hand. I am very sure that this must be the medical method of the future, and a way to avoid all those expensive and uncertain tests.

The big difference between a regular out-of-body experience and the Edgar Cayce readings was the amazing fact that his soul consciousness was at the highest level in the ethereal realm, but still had full contact with his body, which allowed him to speak.

Somehow the *Akashic Records* remind me of Carl Gustav Jung (1875–1961). He had the suspicion that the minds of all people are somehow interconnected; he called it the collective unconsciousness. How right he was!

It is useless to cite examples of the Cayce readings in order to prove reincarnation. This is not necessary because he is 100 per cent proven and documented. But it may be profitable to illustrate just how reliable his readings were.

The readings gave the same information regarding the whereabouts of Jesus from the age of twelve to thirty as given in the records of the monastery in North India. But Cayce did not say anything about Jesus' last words in Mayan. Maybe because nobody asked such a question.

Other readings about Edgar Cayce himself made clear why he was able to do what he did. Even though the man Edgar Cayce was a very simple and rather uneducated man, his soul, the Cayce entity, was highly developed. Edgar Cayce could see the aura of all people during his life. He stopped playing cards because it was not fair to the other players. He always knew exactly what the other players had in their hands, and which card would be the next they would use. The schoolboy Edgar Cayce had a tremendously hard time with his spelling book. An inner voice that came so often to his rescue, told him what to do. He put the book under his head and slept on it. When he woke up, he not only knew the spelling of every word in that book, but even the page it was on.

In some readings he stated that the Great Pyramids were built from 10,490 to 10,390 B.C., more than 12,000 years ago. And his own soul entity was at that time incarnated as Ra-Ta and very much involved in the construction of the pyramids. He supervised their construction which Hermes, a scientific and engineering genius, had designed. He also said that Hermes was an incarnation of the Christ entity, Jesus.

Of course, this last story is very hard to swallow because it is contrary to all the references we have in our books, and therefore just unbelievable. Everybody knows that Cheops built the Great Pyramid, right? Every book in the world says so! Let us see:

There was an Englishman in 1835 who wanted to become famous by making great archaeological discoveries, as was the custom of the English upper class at that time. This British Colonel was the black sheep of a prominent family, which sent him on a Mediterranean cruise. However, he got stuck in Egypt.

His first great discovery was a coffin in the third (small) pyramid, which contained a skeleton. This discovery established who built that pyramid. Much later, the professional Egyptologists found out that the coffin lid was from another dynasty, which had a ruler with the same name.

The skeletal remains are from millennia later, from a Christian burial. So they removed all references to this "great find" from all records.

The so-called Great Pyramid, the largest of the three, was the next project of this clever guy, Colonel Howard Vyse. He had learned that so far it was unknown who built this pyramid. When he saw that there was a flat chamber above the ceiling blocks of the king's chamber (the builders left a passage into it), which had the same kind of ceiling blocks as below, he got an idea: Maybe there are more layers of such reinforcements, resulting in more flat chambers. So Vyse hired an English mason and ordered him to use gunpowder and mason tools to force a way up at the side of the king's chamber.

Altogether they found five extra ceilings and five flat chambers between, each one about two feet high. Then, Howard Vyse provided a drawing, showing the location of "quarry marks" in the upper flat space, and a facsimile of the quarry marks that his assistant, Mr. Hill, had copied. (Note: The last people who were up there were the builders!)

The news was sensational. Archaeologists went up and verified everything. All the material then ended up in the British Museum, and from 1837 onwards, the whole world knew that Cheops built the Great Pyramid because a cartouche in red paint had the insignia of Khufu (Cheops).

Sometime between 1976 and 1980 Zacharia Sitchin researched the events of 1837. He had a very hard time getting permission from the British Museum to see the original "Hill Facsimile" which Mr. Hill, the assistant of Colonel Vyse had made from the red paint "quarry marks" above the king's chamber. When Sitchin unrolled the sheets, he saw at once that he was looking at a fraud, a forgery! Everything was badly misspelled and Khufu's (Cheops) name was written in gross violation of Egyptian religious laws. It contained the symbol for Ra, the supreme god of ancient Egypt. No ancient scribe would have dared do this.

Sitchin included these findings in his book *Stairway to*

*heaven*, and expressed his suspicion, that Mr. Hill produced the quarry marks himself during the night of May 28, 1937. But he had no proof for this suggestion.

Then, in 1983, he got unexpected help. He received a letter from a Mr. W. M. Allen of Pittsburgh, containing a copy of an old letter. This old letter was written by the mason whom Colonel Vyse had hired to force the passage up to the four other flat chambers. The family, who had emigrated from England, still had this letter, written by the great-grandfather in Egypt in 1837. He wrote that he had witnessed Mr. Hill going into the pyramid at night with red paint and brush. When he objected and told Mr. Hill that this was not right, that he was going to commit an archaeological crime, he was fired and banned from the site. Zacharia Sitchin was flabbergasted, because this was proof of his own conclusions beyond his wildest dreams.

Therefore, since 1983 we have three things straight. (They are still not in our reference books, and will not be for a long time, because the Egyptologists do not want to admit that they were fooled for almost 150 years.)

First, the scientists who had difficulty calculating how Cheops built the pyramid do not have to worry anymore. Cheops ruled for only twenty years. If his people worked every single day of these twenty years for ten hours, they would have had to quarry, shape, transport and lift into place one of the two and a half million blocks, weighing two and a half ton each, every *1.7 minutes*! That is simply impossible.

Second, the unmasking of the Great Pyramid forgery must be a great relief for the expert geologists. They have always held that the Great Pyramid and the Sphinx must be much older, at least ten thousand years old. They were sure about that by comparing the erosion with other objects of known age in the world, and under similar climatic conditions.

Third, the readings of Edgar Cayce about the age and history of the Great Pyramids are now believable and make much more sense than the information we get from our encyclopedias, especially the theories of some scientists who explain how the Egyptians built the pyramid with primitive manpower and tools.

But how did the builders do it? I have my own idea about this, and I think we can find part of the answer in Florida. At the Southern end of Florida we can visit the Coral Castle located in Homestead. It is now a State Park, and Hurricane Andrew could not damage it, because this fortress-like complex is constructed of massive coral blocks, many of which exceed five tons. But there are some items that weigh up to twenty tons, and the heaviest single block has thirty tons.

You may think that this massive Coral Castle was built by stone-ragging slaves of an ancient civilization, but this is not the case. The entire complex was built between 1920 and 1940 by and for one man working alone and in secret! His name was Edward Leedskalnin. He emigrated to North America before the outbreak of World War I from Latvia.

At five feet tall, weighing 100 pounds, and in uncertain health, Leedskalnin would be an unlikely candidate to quarry and move these tons of coral. He was a fanatic for secrecy and worked only after sundown, when he was certain no one was watching him. When we consider that he cut, moved and positioned all of the structure's megalithic blocks in the dead of night, the man's achievement assumes an incredible scale. How did he do it?

For only one time he relied on outside help when he moved the whole castle to another location in Homestead. He hired a truck, but insisted that its driver not be present when the blocks were placed on the truck. The driver left the truck there every morning. When he returned in the afternoon the chassis was loaded with coral monoliths. Once, the driver absent-mindedly returned after less than half an hour for his lunch he had forgotten on the seat. He was astounded to see the truck already completely loaded with the multi-ton stones.

Some teenagers spying on him one evening claimed they saw him "float coral blocks through the air like balloons".

They seem to be the only witnesses to the construction mode of Coral Castle.

Leedskalnin passed away in 1953 of malnutrition and kidney failure. Thirty years later, Coral Castle was placed on the National Register of Historic Places. Leedskalnin is quoted as saying, "I have discovered the secrets of the pyramids. I have found out how the Egyptians and the ancient builders in Peru, Yucatan and Asia, with only primitive tools, raised and set in place blocks of stone weighing many tons." The very stones of Coral Castle support his story. At an average of six tons, they are twice the weight of the blocks in Egypt's Great Pyramid at Giza. Carrol A. Lake, a Colonel in the U.S. Army Corps of Engineers, stated, "Leedskalnin proved for all the world to see today that he knew the construction secrets of the ancients."

I think that my guess is right. The answer to the secrets of the pyramids can be found in Florida! So why are the Egyptologist still searching for a solution? And now I will close the circle: The Edgar Cayce readings also stated that the Pyramids were constructed by means of levitation!

If we think about the few examples in this chapter, which are only a tiny fraction of all known cases, then we must be overwhelmed. The only logical conclusion is to accept reincarnation as a fact, and not as fiction.

I must not forget one very important viewpoint. The Bible says in many places that God is just. Then how come so many good people live under terrible conditions or in bad health, when at the same time very bad people live a life of luxury? What is just about people who are born with disabilities and die painful deaths in childhood? Only when we apply reincarnation and the law of karma to these "unfair" conditions does everything make sense.

In Chapter V we discussed the answer to the question of why we have reincarnation on this planet: To grow in character, develop our God-given power of love for other people, and work our way back to our real self, to our heritage, the superconsciousness in the ethereal realm of God.

In one reading, Edgar Cayce responded to the question, "Will I go to heaven?" with this very good correction: "You do not *go* to heaven, you *grow* to heaven! See?"

So let us end this chapter with a quote I found in an ARE publication. It fits not only this chapter, but the whole book:

> I would rather have a mind opened by wonder, than one closed by belief.

# Chapter VII
# THE STUFF OF MIND AND CONSCIOUSNESS

> Nature is not only weirder than we think, but weirder than we can think.
>
> J. B. S. Haldane

All of us have a mind. It is what we think, our free will, our intellect and memory. We also have a consciousness, which is the awareness of our existence, and of what we are.

Where is it located? Most of our scientists are searching for the location of mind and consciousness in our brain. And from their viewpoint, and from the belief of most non-scientists, they have good reason to do so. If somebody hits you over the head with a club, you are unconscious, so the brain must have been affected. Why do we say that we "have it in our head" when we know something? The sensing organs of seeing, hearing, tasting and smelling are all located in our heads, directly connected to our brains.

Signals between brain cells are transmitted by a complex network of microtubule in bundles. The microtubules have specialized protein cells, which work by a very sophisticated system of electron exchange. Anesthetics "freeze" these electrons in place, and make you unconscious. Drugs, like LSD and others, work also on the brain and can cause hallucinations and permanent brain damage. Other drugs can help you when your brain is sick and you are mentally ill. Alzheimer's, which permanently destroys brain cells, changes a person into a mental wreck. All of these facts, and many others, including brain surgery, point clearly to our brain as the home of the mind and consciousness.

For a materialistic scientist this evidence is sufficient. It is the brain, period! But there is an entirely different set of proven facts. So we have these questions: How did the biological brain of Edgar Cayce manage to find out what is going on hundreds of miles away? And worse, how could his brain find out what happened thousands of years ago? How did he "read" and memorize a whole book while asleep?

Where is our memory? Where are our emotions? And what does it mean to have a near-death experience or an astral projection (an out-of-body experience)?

In Chapter VI, we talked about cases of reincarnation, especially the case of Shanti Devi. She did not have the brain from her former incarnation available, but she knew everything. What is this third memory system that knows all, and connects two different incarnations?

Because all of these findings are also proven facts, as we have already seen, there seems to be a lot more to know about the workings of our mind than only the brain. It must have something to do with the fact that we are in reality spiritual beings, and not made of physical matter only.

Our brain definitely has much to do with our mind and consciousness, but it is not the seat of it. The brain does have its own mind, but only for immediate use in connection with the senses. The main function is to be a sophisticated switch system between the body and the soul, working in both directions.

But first we have to take a look at our knowledge of the brain. In one decade our researchers have found more facts about the function of the brain than during the whole of the last century. (This applies actually to all branches of science and technology.) The people who work in brain research could be split into two groups, the materialists and the philosophers.

The materialistic-oriented scientists are sure that the brain is a sophisticated biological machine and that mind and consciousness are only a phenomenon, an illusion. They are also the ones who believe that some day we will have computers with their own minds, so called artificial intelligence.

In contrast, the philosophers tell the materialists that they will believe in such computers when they see one that gets sad and cries, that can love and hate, and that can appreciate and enjoy beauty and the smell of flowers. These scientists do not believe that the brain is all that is there. The philosophers have a very good point here, but our society has become so extremely materialistic during the last century, it is no wonder that the materialists are declared the winners, at least for now.

The people of both groups are neuroscientists who have really made much progress in understanding the brain so far. Until now, at least two to five per cent of the secrets of the brain are known. Most of the secrets are still unknown. I will give a very brief picture of what we know so far, or think we know.

First, the tools our scientists are using. The latest and one of the best is the MRI, the magnetic resonance imaging system. It produces cross-section pictures of slices through the brain. PET, positron emission tomography, tracks the blood flow in the brain. The strange SQID, superconducting quantum interference device, picks up magnetic field marks of brain action. Then there is single-photon emission computerized tomography, which also tracks the blood flow, but in a different way than PET. And finally the EEG, the "old" electroencephalogram, known as the brain scanner, detects the electrical brain activities. These are all relatively new tools for research and mainly responsible for the fast progress in brain research. All work on the living brain without any kind of damage to the person. And now we take a look at the brain.

I can give only a very rough picture of the brain, and only the most important parts. The anatomy of the brain is much more complex than I can explain here. (See Figure 2.)

Where the spine enters the brain, we have the brainstem, which controls most of the automatic (motor) body functions, like breathing, etc. (Einstein joked about this, saying that the brainstem would be all that is required for soldiers to march along. He was a pacifist and disliked the military maybe more than anyone else.) The brain itself has many different sections. The cerebrum makes up two-third of the entire brain and has a deeply convoluted surface; it is divided into two interconnected halves or hemispheres. Each of the halves is divided into the frontal lobe, the parietal lobe (middle), and the occipital lobe (rear). The outer layer, about a quarter-inch thick, is called the cerebral cortex; it is gray and seems to perform the functions of mind and memory. In the rear, above the brainstem, is a small extra brain, the cerebellum. It is the command center for all functions of the body. In the center are many specialized parts, all of which are relatively small. The pituitary gland appears to be the most important of these small units, because it commands the whole hormone system of the body, which is extremely complex. Generally the brain seems to be organized in this way: The gray cerebral cortex does the work, and the inner white mass is the "wiring" of the brain, interconnecting everything.

The human brain contains about one trillion cells; 100 billion of them are neurons. Neurons are very complex cells and there are dozens of different "models" of them. Each model has an entirely different set of special functions. They do not work in a binary system, like our computers, which have only on or off for each transistor, one byte at a time. The neurons are superior to that. Each one is interconnected to others by dozens, even hundreds, of sylph-like fingers, called axons. This allows the kind of high efficiency computing that our computer designers can only dream about. This parallel computing capability stretches the capacity of each neuron by magnitudes. Hundreds of thousands are working right now as you read this sentence. The letters are automatically realized and the words are formed; at the same time you know what each word means. And at the end of a sentence you know something new or recognize it as something you already know. At the same time you decide to believe it or not, like it or dislike. All of this is too much for our best computers, but every child learns this in school and while playing.

The brain is organized in a weird way. If you ask an intelligent volunteer with a high IQ a tricky question, the brain activity measures very low. The same question given to a low IQ volunteer results in high activity – just the opposite of what one may expect. Intelligence seems to mean a brain-knowledge of what parts of the brain not to use to think efficiently. There is another paradox. While a person learns a certain task, the corresponding area of the brain "lights up" more and more and the spot enlarges. As soon as the task is mastered, this spot will not light up again when the task is repeated. So far scientists have not discovered where the brain stores the newly learned skill. (Later we will discuss where it goes.) Generally, most procedures of the brain end up somewhere in the cerebral cortex, and the memory seems to be stored there. Because a human can know so much more than the cerebral cortex may be able to hold, many scientists believe that the whole brain is holding memory.

The brain is a mass weighing about three pounds. Almost one-third is used entirely for the function of seeing, and evaluating and interpreting what we see. Other big chunks are used to do the same for hearing and smelling. Only a relatively small portion of the brain is left for the higher mental functions of the mind, maybe about 30 billion neurons with at least 500 billion interconnections from neuron to neuron and across the brain. What can we do with this capacity?

Let us take an hour out of a normal day, while keeping in mind what our brain is doing: We enter our house and walk through different rooms and a hallway to the family room. We know our way, we see everything and know what it is, we even know where and when we got most items. Then we sit down at a table and work for a while on a two-thousand-piece jigsaw

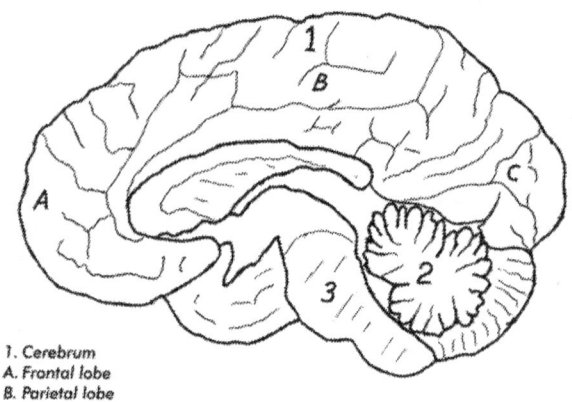

1. Cerebrum
A. Frontal lobe
B. Parietal lobe
C. Occipital lobe
2. Cerebellum
3. Brainstem

A NEURON

NUCLEUS
CELL BODY
AXON

*FIGURE II: THE HUMAN BRAIN*

puzzle. After about twenty minutes we recognize the time and turn on the TV. Now we watch and fully digest thirty minutes of news. While we do this, we agree, disagree, get mad, and have other emotions. Then we read the front page of the newspaper, do not like the news, and go out to the patio.

Every computer and robot expert will agree that all the supercomputers in the world, connected together, could not perform the zillions of computing steps we just did during this simple example. The distributed parallel processing of our brain (and something else, as we will see) is superior by magnitudes. (According to robot designers, the best autonome robots so far are mentally above a slug, but below a fly.)

In *20/20* they showed an interesting man on TV. I do not remember his name. He is a case of a savant, a person who has difficulty with the activities of daily life but has extraordinary memory skills. He managed to read the two-inch thick telephone book of his city from cover to cover, and likewise whole encyclopedias and many reference books of facts. Incredibly, he has retained everything in his memory. He was in a restaurant and read the name of the waitress who served him. Then he said that her phone number was such and such. She was surprised and wanted to know how he knew it. She did not believe him when he explained that he had read it in the telephone book. You can give him any phone number and immediately he will tell you the name and address. Give him name and address and he will tell you the phone number. Ask him anything from an encyclopedia and he has the answer. He can also give the day of the week for a certain date hundreds of years ago. He can also tell you what happened on that date. At the end of this *20/20* TV report (early nineties) he was shown going from school to school and answering every question the students had for him.

How many memory bits are required for such a performance? Quadrillions? Quintillions? I have no idea, so let us use the term zillions. That is for memory. Another zillion neuron connections are required to access each fact and process it, *if* neurons do it! I think that this man's capabilities surpassed the natural capacity of the brain by magnitudes. He stored and processed the information somewhere else.

In the thirties there was a professor of mathematics in Germany. You could give him two six-digit odd numbers and within seconds he would tell you the result of multiplying the two, always correct. When asked about it, he said that he did not know how he did it. All he could say was that he hears the two numbers and "something" starts working in him without him being aware of it, and the result of the multiplication pops up in his mind. I doubt if this was done only by the brain.

Srinivasa Ramanujan is considered the greatest genius in mathematics. He was for mathematics what Newton was for physics and Mozart for music. He lived in India completely isolated from the western world of mathematics. His formal education was basic arithmetic. At the age of ten (the same age another great one, Gauss, was discovered) Ramanujan began to play with numbers, and within a few years he developed the higher mathematics of the last hundred years in Europe entirely on his own, without studying any books on mathematics! Then he stumbled on a book of mathematics and wondered what he did. While working as a low-level clerk, he mailed letters with some of his own theorems to mathematicians in Cambridge (U.K.) in 1914.

All the mathematics professors, except Godfrey Hardy, doubted Ramanujan's authenticity and laughed about his letters. But Godfrey Hardy saw that he had the letter of a genius in his hand. Hardy arranged for Ramanujan to move to Cambridge, where he made great contributions to the world of the highest mathematics. But just like Mozart, he died too soon, at the age of thirty-three in 1920.

Ramanujan's abilities could be explained as a very strong carry-over from another life. The topic of this chapter prompts me to tell the following story about him: While Ramanujan was lying ill with tuberculosis; Godfrey Hardy went to visit him. He used a taxicab having the number plate 1729. He remarked to Ramanujan that he hoped that this number 1729 was not a number of unfavorable omen. "No," said Ramanujan right away, "1729 is a very interesting number. It is the smallest number expressible as the sum of two cubes in two different ways." This is a very complex theorem that would require a modern supercomputer to prove! How did he produce this answer just like that? His brain could never do that on its own, there must be more behind it.

There are hundreds more of such "strange" cases. Let us enjoy two more. In the mid-eighties *Time* magazine published an article about a three-year-old boy in Korea. The name was not given for reasons of privacy. When his father came home, the boy told him that he wanted to become a physicist like his father, because he had read one of his father's physics books lying on the table. His father told him that that was not possible because the book was in German, and he, the boy, spoke only Korean. Now the boy took the book and read out of it in German, explaining that he understood everything. He had never formally learned a single word of German. How does that work? When the boy entered high school at the age of seven, his parents attained a media blackout in order to preserve their privacy because their home was overrun by Korean and U.S. reporters.

Another extraordinary case occurred here in the States. On June 5, 1994, Michael Kearney graduated from the University of South Alabama with a bachelor's degree in anthropology, with honors. So what? What is so special? Michael was ten years old! This is a truly amazing achievement. How did it happen? At the age of eight months he began to read and to speak in complete sentences. "What is for dinner tonight?" was one of his first complete sentences. His mother answered, "Breast milk!" The media (*20/20* and TV news) said that nobody has an explanation for this. That is wrong! Millions of people around the world can explain this case as another case of reincarnation. When I tuned in to a San Francisco TV channel, I was lucky to see the graduation ceremony at

Berkeley (summer 1998). And there he was again: Michael Kearney, now fourteen, received his Master of Science in Biochemistry and, of course, a standing ovation. (I wondered where he was in the past four years.)

Now I am ready to present my view and definition of mind and consciousness. I like to compare the brain to a portable computer that serves us during the day and records everything in its memory. When we are back at home, we connect the portable to our main personal computer. We then transfer everything into the memory of the personal computer.

Our brain works similarly, but there is a big difference. The connection to our main computer (The mind of the soul!) is permanent and all data are transferred as soon as they are processed by the brain. Another big difference is a permanent feedback from the soul-mind, enabling the brain to perform tasks that are beyond its own capacity.

Our brain can do marvelous things on its own; we may call it the physical mind, but most expressions of the mind are not of the brain. It is like a speaker on the wall. The speaker produces music, but the real source of the music is the radio it is connected to, and the radio station which broadcasts the music to the radio. It is a permanent interplay between brain-mind (speaker), subconscious mind (radio), and the superconscious mind (broadcasting radio station). And because our consciousness is stuck in our physical body, we identify ourselves with the body, which is an illusion.

Our seven main glands (the gonads, leydig, adrenal, thymus, thyroid, pituitary and pineal) are saturated with spiritual matter. Of course, matter is not the best term for these spiritual energy units, which are infinitely small. The so-called Cambridge Club in England coined a very good expression in the 1930s. These scientists stated that the stuff of the universe is mind-stuff. It is the stuff of our astral mind (subconsciousness) that saturates our glands and the whole brain while occupying and controlling our body during an incarnation. (We cannot detect these spiritual units with current means of testing.)

There, in these higher parts of us, we find our character, morals, love, emotions, all the real thinking processes, and a practically unlimited memory of data from this and past lives; it is our long-term memory. The short-term memory is still very much a function of our brain, and holds information before it is transferred into the permanent "big one".

When the body and the brain get old, we begin to forget what we had to say, even in mid-sentence. Or, when standing in another room, we forgot why we went there. In contrast, long-term memory remains as clear as ever. Old people can forget new things in less than a second, but they can talk forever about every detail of their childhood. It is said that we cannot take anything with us when we die. This is correct in terms of material things, which are of no importance. But we take with us everything we have learned, and the memories of everything we did! Most of the time, the next life will start at the same level that this one ended. So we better work towards a good finish for our mind, and do not care too much about material things we own, because they are only borrowed for this life.

We read again and again that scientists have found the location of certain mind functions in the brain. This may be correct if the tests were performed for any kind of body movements, which are controlled by the brain. But if the tests were made for any kind of thinking processes or emotions then it is different. What they have really found was the switching area for these functions. They found the spot where the contact with the soul-mind occurs. Or they found the area where certain functions are processed, because this area "lights up" while individuals are learning a task. But then, as I said a while ago, after mastering that task, nothing lights up anymore when the task is repeated. So where did the brain store it away? The data was transferred to the "big one" and from then on it is an automatic skill. (And we take it with us! That is why some children who have never had a lesson, sit down and play skillfully, the piano, for example.)

Mental illnesses, including Alzheimer's, are only symptoms of a bad connection between soul and body, a malfunction of the switchboard brain! All mental diseases are in effect caused by the brain, because the switchboard brain is sick. Therefore the contact between soul and body is bad, or even completely interrupted.

It is often possible to help a mentally ill person by doing something to the brain. Medication or ECT (electroconvulsive therapy) may reconnect the soul with the body. Surgery may help to correct a malfunction of the brain. Very often the brain will rewire itself after such surgery and suddenly other areas act as the switching points. I think the brain uses its own "mind" to do this.

Like everything else in this world, not all of these body-to-soul connections are equal. When we are born, the brain is prewired for two conditions. First, it is completely programmed for all vital body functions. Second, it is programmed and has all the necessary connections to the mind of the soul, so it will know what to do and how to react to its surroundings. Very few mothers on this earth know that a baby has much knowledge i.e., the knowledge of the soul. That is why a newborn can outsmart and "train" the mother. You can test this. If you meet a baby of no more than a few months, do not talk, just "think" to the baby: "I know that your mother does not know that you know so much. Is not that a lot of fun?" The baby will smile at you in a sly way you never expected.

But this brain-soul connection does not last forever. In some babies it is completely lost after a year, for many others it lasts for three or even five years. Then a full amnesia of the soul memory sets in. And here the danger begins. During this time it is of utmost importance that the child's mother or primary caretaker keeps the child very busy with all kinds of entertainment, because the brain has to learn! These first years, three in average, are the most important years for the future intelligence of this person. The inequality of the brain-soul connections begins already in these important years of our life. (This explains also why most of us cannot remember anything before the age of three. The consciousness was still with the soul at that time.)

Generally, the inequality starts with the fact that the chances or opportunities for learning (programming the brain), are not the same for all people. As a result, the quality of the contact system (brain) is very different from person to person. Therefore, when a person's connection to the soul-mind is very bad, we tend to classify her or him as dense or stupid. Other brains are exceptionally well connected to the big "soul computer" and are classified as very intelligent, or having genius. (Like the telephone book guy; or the genius in mathematics.) Others are somewhat too well connected to the long-term soul memory, much more than they should be, like Shanti Devi or the little Korean boy. This latter connection is normally supposed to be completely blocked (total amnesia), so we will learn by heart in this life, and not by calculation. This amnesia can be removed in almost everyone of us under hypnosis. (A warning: If you go to a doctor for hypnosis in order to learn about your past, be very careful! Many of these people are not honest and make suggestions under hypnosis, resulting in stories that are not true.)

What is the fate of our consciousness? What happens after we die? The complete answers to these questions, which are man's eternal quest, would make for a very long book. So all I can do here is to give a short rundown of what I know so far.

When we make the transition (die), our consciousness moves back to the level that is right for us; according to our karma – in most cases to the subconscious mind of the astral level. But a few go all the way through to the ethereal level of the superconscious mind for different reasons. Most do so for greater learning potential because they are already far advanced in their development. Others want, or have to go, to other planets for a while. A few remain in the etheric for good, because they have made it and do not have to reincarnate anymore. Then they are again co-creators of the causal cosmic intelligence and capable of understanding the purpose and properties of the universe. For them it is the finish of a long journey, a finish everyone of us will eventually reach – an ultimate state of enlightenment, where everything is possible. Right now, while still in the physical world, we can only guess the purpose of our lives.

Most of us go to the so-called astral realm after death. So, what happens when we arrive there? The very first thing we have to face is what has been described in near-death experiences. We float through that tunnel, meet the light being, have a rerun of our life, and are judged – mainly by our own higher self – about the life we just lived. And then it becomes different than a NDE. We meet entities we know. One of them we recognize as the guiding entity we had during our last life (a guardian angel for most people). This very old friend takes care of us from then on. We also recognize that everybody is only a point of energy and light in an array of colors. White for the new ones, and all the way through colors of ever-higher frequencies to violet, the color of the souls that have advanced the most in this realm. All communication is by thought; everyone can read your mind, and you can read everyone's mind. However, you can block your mind against unwanted reading.

Our guide leads us through a cleaning process, a kind of spiritual shower. As a result, our mind is free of any hangover from the life we just lived. Now we are ready for our final destination. We enter a room that is not a room in the sense of a three-dimensional room. There we meet a group of souls that we know. They are not from the earth, but rather this new realm. Some groups consist of twenty members, some as big as fifty. Immediately we know that we *are back home*! Our life on earth was just another trip. We may even miss some entities and are told that they are incarnated at the moment.

Our time in this realm is primarily spent learning. Time as

we know it does not exist, but all things happen in a sequence. After a while we are ready for the next incarnation; this can happen after a few earth-months, or after thousands of earth-years. It is different for every soul.

For more details about our life in the astral, and the most comprehensive picture, read the book *Journey of Souls* (listed in *Suggested Reading*), and get as many books about Edgar Cayce as you can. Then combine everything in your mind and you get the complete picture.

After some preparation, we begin to plan the next life on earth – where will we live, who will our parents be, and what incidental opportunities or tests will await us, and many other details. With others in our group we determine how we will live together. Some will be family members, others friends, and some just an acquaintance. All of this is programmed into our subconscious mind. Then we move into the physical realm and take care of the development of the chosen fetus without attaching to it. The final occupation, the actual incarnation, takes place at the earliest six weeks before birth. Most souls prefer to enter just after birth because they know that the birth procedure hurts. (According to Edgar Cayce.)

And then we live the next life. We may follow our plans for this life, by listening to our "inner voice" – the subconscious mind – and make spiritual progress. Or we may ignore the inner voice, do the wrong things and fail. It depends entirely on our free will to make this next life a success or a failure.

All brains on earth are different in their capabilities, as we know. The minds of the soul, the superconscious minds, are also not all equal. That is why the "smart" ones are at the top of the ladder of the universal hierarchy, in charge of solar systems, galaxies, and galaxy clusters. (The archangels?)

Why do we and most animals have to sleep? A huge amount of spiritual energy is required and used up by the activities of the day. During sleep we recharge our batteries through our glands. We can also do a lot of recharging in meditation. (Do cats meditate? Sometimes I wonder.)

There is another interesting fact. The larger animals (and brains) can depend very much on their own mind and consciousness. Smaller animals depend more on spiritual energy and guidance than larger ones. Otherwise, they could not do what they do. A very small fly, less than an inch long, flies around our living room for a whole day. Where is the fuel tank that permits such a performance? A hummingbird in the South – which, over the summer, collects less than half a gram of fat (only four calories) – can fly across the Gulf of Mexico. These and thousands of other tiny creatures depend very much on spiritual energy for fuel.

Very few readers will believe what I have just stated. Spiritual fuel? What is that? Never heard of such stuff! But before you make up your mind and reject this, let me tell you one very well-known example of transforming spiritual energy into mechanical energy and physical matter. This case is 100 per cent tested and documented. It is officially notarized by the German government and the Vatican. If you should get interested in this case (and many others), look for the literature that deals with it; it is fascinating. Here I will only mention the details that are important for the problem at hand – spiritual energy and fuel.

It pertains to one Therese Neumann (1898–1962). She lived in Konnersreuth, Germany, all of her life. On Easter day 1926 she became stigmatized. She bled heavily out of the palms of her hands, on her feet and her forehead. In short, like so many other people before and after her, she had the same wounds that Jesus had on the cross. From then on, this happened to her every Friday. Each Friday, she lost some weight because of the bleeding and stress, but always regained it until the next Friday. This on its own would be understandable, but on this Easter day in 1926 she also stopped eating completely! Until the end of her life she drank only a little water.

Of course, she was tested more than once under strict scientific conditions and government control. She was always a very happy and healthy woman, except for being permanently bothered by the media and scientists. Being a Catholic, she asked the church for help. So the bishop of Bavaria issued a decree of privacy for her. From then on, his permit was needed to visit her.

There are other people on this planet who do not eat at all; but this case of Therese Neumann is the most famous and best documented one. Of course, what Therese did was, and still is, called a miracle; but we know already that miracles do not exist. So, what did Therese Neumann do without knowing how? She used and transformed spiritual energy into matter in order to regain her weight. She also transformed spiritual energy into known energies to run her body. Of course, Therese never knew what she did, not in this life.

What Therese did should be of no surprise, because there is a law in physics about the preservation of energy. Energy cannot be destroyed, only transformed into another form of energy or into matter. The only problem with the case of Therese Neumann is the fact that "spiritual energy" is not in our physics books because so far science is ignorant of it. We do not know how to handle this energy, or how to develop mathematical formulas for it. Therefore everybody talks about such cases in terms of a miracle.

Let us see what else can happen when spiritual energy is transformed into physical matter. Another well documented "miracle" is this case:

In 1868, Pierre de Rudder, a Belgian peasant, fell from a tree and broke his leg so badly that a piece of bone, over two inch long, came out of his leg and had to be removed. The lower part of his leg dangled freely, held in place only by skin and muscle. The doctors wanted to amputate the incurable leg because walking, or even standing, was impossible. At that time they had no means of repairing it with surgery. But de Rudder refused and lived for eight years in terrible pain, unable to walk. Then he decided to make a pilgrimage to Oostacker, where a statue of "Our Lady of Lourdes" existed.

Three helpers were with him to move or lift him as required. Sitting in front of the statue he prayed earnestly for his wife and children.

Suddenly he was overcome by a fit of ecstasy. He stood up without his crutches and walked towards the statue. He was instantly cured on the spot!

When he died in 1898, an autopsy revealed that a new piece of healthy bone had formed between the ends of the bones. Pierre de Rudder's leg bone can still be inspected at the University of Louvain in Belgium. (This story and many more are in the well-researched book *Miracles, a scientific exploration of wondrous phenomena*, by D. Scott Rogo.)

Of course, this was a miracle. What else? And according to the Vatican it happened through divine intervention. But it was not a miracle, nor divine. It was just more proof of the tremendous creative power of the spiritual thought energy. What he did was connect to his own superconscious mind in deep prayer, resulting in instantaneous action by the ethereal part of him. Without knowing it, he applied the same principle that Jesus used for his "wonders".

Of course, you can develop your own ideas about this case. Did Jesus do it for him? Was it God Himself? Was it Mary? Actually it does not matter, because it is all the same thing, all from the same source!

Before we change to the last topic of this chapter, let me remind you of the strangest case of them all. You can read about this man in every good encyclopedia. He lived from 1773 to 1829. He could read at age two, and by age four had read the Bible twice. By age twenty he knew Arabic, Hebrew, Persian, Turkish, and seven more languages besides English. So in 1801, when he was twenty-eight, Thomas Young was already a professor of natural philosophy at the "Royal Academia" and lecturing on acoustics, optics, gravitation, astronomy, tides, the nature of heat, electricity, vegetation, animal life, climate, cohesion and capillary attraction of liquid, the hydrodynamics of canals, harbors and reservoirs, techniques of measurement, common forms of air and water pumps, and new ideas on energy. Enough? Not for this guy!

He published a new theory about light being a waveform. He also conducted a very famous experiment by sending light through adjacent holes, producing the now well-known interference pattern. Then he expressed his theory (since proven true) that the retina of our eye receives only three primary colors, and that the brain produces all the colors out of them.

Then he seemed to get bored and he turned his interest to something else. He began to crack the secrets of the Egyptian hieroglyphics using the Rosetta Stone in the British Museum. On this stone tablet are the hieroglyphics right beside ancient Greek letters.

When I think of genius I do not think of Einstein, Newton, or any other person. The name Thomas Young enters my mind, and for a reason – he was hundredfold proof that the cell-capacity of the brain can be exceeded by magnitudes because of the spiritual connections to the higher levels of our own existence.

## About Prophecy

Our topic of mind and consciousness leads us to another question that is very much in fashion these days: What is prophecy? How does prophecy work? Is prophecy possible at all, and if so, how reliable can it be?

There are prophecies in magazines, in TV-specials and, of course, much talk is going on (and always was) about the prophets and their prophecies in the Bible, and about Nostradamus, the French doctor and astrologer (1503–66). There is a lot of excitement about prophecy.

People have all kind of questions about the future: What will happen in the year 2000? Will California go into the ocean? Will there be a Third World War? Will the president have a divorce while in office? Will we have a pole shift, and when?

Because the topic of this chapter is mind and consciousness, we will not talk about individual prophecies. Instead we will investigate the question of *how* prophecy works.

It is a fact that many prophecies prove to be right. It is also a fact that most prophecies are wrong. But let us reason it out. If only a few would be right, then we have good reason to find out why and how, because true prophecies do exist. And while doing this, we may also find out why so many prophecies prove to be wrong. (I only talk about real prophecies, not about tabloids and telephone psychics.)

We have already seen that non-physical realms do not have time as we know it or as we are used to it. Our kind of time (linear flow) was created as a mandatory prerequisite for the existence of the world of matter and related energies, the physical universe. (The same applies for space.) As long as we live in this physical world it is absolutely impossible to look ahead in time, not even for a second. Time approaches us from the future, second after second, and we have to take and accept what it will bring us.

Just a few pages ago we have seen that the connections of the brain with the soul-entity vary from one person to another. We have to keep this fact in mind, because it has very much to do with the conditions for prophecy.

Some people have the ability, or gift (or the better connection), to enter the ethereal realm in trance, dreams, or otherwise subconsciously. In doing so, their consciousness enters a level of existence without time as we know it. It is not easy to adjust to this condition if you are not used to it. Everything that happened in the past or will happen in the future is recorded there as *now*! (Why and how will be clearly explained in Chapter IX.) This does not mean that everything in our future is already fixed and predetermined and we cannot do anything about it. Not at all! We have our free will, and we use it to form the future. Of course, we cannot change the happenings in nature – earthquakes, storms, and other disasters.

The *Akashic Records* register everything as *now*, but in the right sequence. This means that everything from the past and future is recorded, even though the future will be formed by our free will. The *Akashic Records* just know everything in advance, so to say. Every entity who lives in this realm of permanent now without linear moving time can look at the past, present, and future in one view at the same instance, but in the right sequence.

And this is the great obstacle for all prophets: The sequence without time! In our world, sequence means that all happenings occur one after another, nicely connected by a moving, measurable time. The mind of the prophet who enters this realm of sequence without time is not used to this condition and gets easily confused. And this confusion is the reason for the differences of the accuracy of the prophets, because the minds of these souls are different. Some are totally confused and mix everything up. Others see a little of the future, but very cloudily, while others see it more and more clearly. A few highly developed souls are able to see very much and clearly.

A sequence without a time makes it very hard, if not almost impossible, to place a future event at a certain date. Now it becomes clear why predictions for only a short time ahead are the most accurate. The farther ahead into the future a prediction has to be placed, the less accurate the date for this prediction will be.

Very few people (Edgar Cayce, some of the biblical prophets, and maybe Nostradamus) have felt at home in this realm and could read the *Akashic Records* very well. The reason for this was that these souls were already much more highly developed, and therefore felt more at home in this realm. As a result, their predictions were more accurate in respect to the timing of an event.

But even Edgar Cayce, and maybe the others too, fell sometimes into this sequence/time trap, and had the timing of a prediction wrong, even though the content of the prophecy was right. In the case of Edgar Cayce, I think I know why. Most of his readings were for a certain person (entity), and his soul-mind was so much preoccupied with the life story of the entity, that the extra bonus, the fallout of a prediction, fell short in respect to an exact timing of the event. The few readings that were done for the sole purpose of prophecy came out much better at the accurate dating of an event.

Therefore, when you see or hear any predictions, please consider the fact that the date for an event is always the most critical part of a prophecy. Most of the prophecies we hear these days state only the event, but no exact date for the occurrence of it, because of the reasons I have just explained. And beware: If a source has a very good record at short-term predictions, this is no guarantee that long-term predictions will be equally correct.

I am very sure that the few predictions by Edgar Cayce, that did not come to pass at the predicted time, do not have to be wrong in respect to the event. They will come to pass, but when? I guess we will never know what misled his soul into incorrectly dating an event.

Maybe you were under the impression that I have studied this field thoroughly, and therefore you would like to know what I thought would happen at the turn of the century, at the beginning of the year 2001 (which is the first year of the next century, not 2000). My answer is: *nothing*!

Why did I believe that nothing would happen? I did so for many good reasons: There are the phony psychics who operate as doomsday prophets. They printed monthly newsletters, or even magazines, with all the bad news about the total disasters that awaited us at the end of this millennium. They charged prices for their stuff, which were very competitive with trial lawyers. What did they need all the money for, if they believed in their own prophecies? While terrified followers lived in caves, and waited for doomsday, these people laughed all the way to the bank.

The historic evidence in the Gospels and the facts of written Roman history have proven that Jesus was actually born in the year 6 B.C. Our incorrect dating system is based on medieval Christian theologian's decisions, which were wrong. Therefore the real Christian year 2000 was 1994!

But it will be the year 2000 only in Christian culture. For the Jewish people it is 5758. For the world of Islam it will be 1421. The Hindu year will be 2012; the Japanese talk about 2615, and the Confucian year will be 5637. The Christian year 2000 is not global! Many more non-Christian people live on earth than Christians. (And God loves them all!) Anyone of these numbers is an arbitrary date without any meaning for the history of the earth. For the majority of the people on earth our year 2000 means nothing.

Only two prophecies may be of a certain importance. Of these, one scheduled in the months of July or more probably September of 1999 has already passed without any untoward incident. Nostradamus had some bad news for this period. Something terrible was supposed to appear from the sky. The other one is December 25, 2012. For the Mayan calendar this is the beginning of a new age. It says nothing about what will happen. (I think nothing will happen, because on this day there will be a total solar eclipse, and the Mayas had their time very much oriented by eclipses. So, it is another arbitrary date.)

Many people read a lot of prophecy out of the Revelation in the Bible. They should read the book *Revelation* from the ARE Press (Association for Research and Enlightenment, the Edgar Cayce foundation). This book is a commentary, based on a study of twenty-four psychic discourses by Edgar Cayce. This book *is* a revelation! It shows clearly that the revelation is a coded instruction for deep meditation of a master. Coded in such a strong way, that nobody could guess and translate wrong. The code can be broken only in a psychic way, not by logic, not with knowledge. Edgar Cayce broke the code; it is fascinating. Revelation has nothing to do with prophecy!

There was also a scare about a "dangerous" planetary conjunction on May 5, 2000. There were much better alignments of the planets in the past, as on February 4, 1962, and nothing happened. There is even a book on the market (390 pages) that has the title *"5/5/2000", Ice: the Ultimate Disaster*. It is very scary!

So, I hope you have not sold off your homes to head for the hills.

# Chapter VIII
# SPACESHIP EARTH

> The great tragedy of science – the slaying of a beautiful hypothesis by an ugly fact.
>
> Thomas Henry Huxley

The earth is our home. Actually it is nothing more than a big, self-contained spaceship in orbit around the sun. As we already know, it represents less than a tiny speck of dust within the universe, but for us it is still very big. If we want to take good care of our home in space, and we must do so, we have to know why it is the way it is, and how it works. It also helps to know what to expect, what dangers are hidden, and how environmental disaster can be prevented. To find out what happened in the past can also be very helpful for controlling our future.

We know that the earth is self-contained. What we cannot see easily is the fact that this ability of self-containment rests on a very sensitive balance of thousands of factors. If any one of these factors gets too far out of the great harmony, a chain reaction of events may be started and disaster can strike. Luckily the self-healing capacity of the earth is very good, but it can be pushed only so far; it has a limit!

There is not only the well-known balance in nature between plants and animals; there is also a balance of planetary factors, which provide the proper foundation for nature.

Size-wise the biggest factor for the possibility of life is our distance from the sun – not too close to be scorched (like Venus), and not too far away to be too cold (like Mars). The next factor of great importance is the angle of the rotational axis to the ecliptic, which is the plane of the earth's orbit. With

23½° it is just right to cause the needed seasons, the ocean currents, and the weather. We have a tropic that is just going strong enough, but also enough ice at the poles for the control of the climate and the ocean levels. These ice caps fulfill a very complicated job of being a buffer for weather extremes. The twenty-four hours for one rotation of the earth are just perfect for the kind of life we have on this planet. The size is also perfect, large enough to hold a permanent atmosphere, but not too large to crush us with a high gravitation and atmospheric pressure.

Here I would like to interrupt: As you can see, the book is getting more scientific. Please let me remind you of the introduction. This is not a book about reincarnation, nor a book about astronomy, but as the title says, it is a book about *everything*. The whole content of this book is required to get the right picture of the universe and ourselves.

The atmosphere of our earth is not too dense, but thick enough to protect us from the dangers of space (UV-rays, X-rays, and a rain of meteoroids). The 78 per cent nitrogen and 20 per cent oxygen plus some water vapor is just the right mixture for the support of our form of life. And this mixture is also perfectly tuned to the gravitation of our planet, to serve the respiration systems of all animal life. This pressure is very critical for the pleural cavities of land animals, so the lungs will not collapse. Because of their structure, lungs cannot be bound to the chest cage. A vacuum between the lungs and chest cage does the trick when we inhale. Venus is a very good example why we are lucky with our air mixture. Venus is the same size as the earth, but the atmosphere is a composition of such heavy compounds that even our space probes were crushed under the horrible pressure when they landed. (Venus is hell with this pressure and the 600°F temperature.)

The ratio between the surface area of the land masses, oceans, and polar caps is another life preserving factor. The oceans are large enough to evaporate enough water for rain, and the polar ice caps hold the surplus water to keep the sea level constant. There are many more things that are in the right proportion to each other; the forests, the mountains with their glaciers and snow caps, the deserts, lakes, and rivers. All these factors, and hundreds more, are essential for nature to survive.

Even the moon plays a role in the dynamics of ocean currents and weather. Life had to adjust to the conditions caused by the moon, because the influence of the moon is a negative one. The gravitational force of the moon causes too much stress on the earth's crust. It causes not only the tides of the oceans, but also a tide of the solid crust of the earth, triggering earthquakes and other disasters. Actually, we have to look forward to the day when the moon finally gets caught by the sun's gravity, after spiraling away from the earth, now at the rate of more than one inch per year. (This process will need millions of years.)

If we look at this fantastic balance, which is really too perfect, then we get the impression that this could not happen by chance or accident. But many scientists believe that we could be the only planet in the universe with carbon-oxygen based life because the chances of it occurring on another planet are nil, in their opinion.

Looking at all of this with an open mind, we begin to believe the many sources who state that the earth was manipulated and formed as required, and is still not finished. (But we may finish it off!) Our planet was developed to harbor carbon-oxygen based life. Life based on other elements is thinkable and theoretically possible. Maybe there are planets with silicon – or other elements – based life out there. Some day we may find out. (We have such life here on earth. Many creatures live in sulfur lakes and other "impossible" conditions for life, without carbon or oxygen.)

Because our earth is only 4.5 billion years old, but the cosmos existed already for an eternity, it was not necessary to design and develop every required plant or animal each time the climate or ecology changed, calling for a line of different plants and animals. Almost all were just carried in from other planets in our or other galaxies. (I will discuss how this occurred in the next two chapters.) First, the smallest and simplest life forms were brought to the young earth for the biological preconditioning of the planet. Then more and more sophisticated plants and animals arrived, perfectly matched for the climate of the time. Through reasoning we learn that life on earth must have been seeded in such a way.

Here is one example: For over 150 million years we had almost no other animals but reptiles on earth. Then came the mammals. Why? The earth was too warm or super-tropical at that time. Since reptiles are cold-blooded the climate suited their inherent nature. Even the huge monster dinosaurs were happy because they did not have the heat dissipation problems of our largest mammals today. But for mammals the climate in these 150 million years, from 220 to about 70 million years ago, was simply too hot. Of course, that is my own opinion; I did not read or hear it. But I arrived at this conclusion after studying most of the available literature about this topic. It is the only thing that makes sense to me.

Not only our earth, but also the whole solar system is properly designed and arranged, and includes a number of features that ensures a continued existence of life on earth. The most important of these features is a buffer zone of four huge outer planets (Jupiter, Saturn, Uranus and Neptune), each one with many moons. They have the job of protecting the four inner planets (Mercury, Venus, Earth and Mars) from incoming space debris, rocks, comets, and big meteors. They collect them with their powerful gravitational fields and so protect the earth. The many moons are a great help in catching even more.

But it is not easy to be perfect in a rough and dangerous universe. We had bad luck too, and now permanently have to face the consequences. There is an unoccupied stable orbit around our sun. Newton said that there could be another planet, according to his calculations. In his time asteroids were still unknown. But where now there are asteroids to fill the space between Mars and Jupiter, there once was a planet! Many independent sources say that the name was Maldec. Some spiritual channels insist that Maldec was destroyed through misuse of nuclear power by the people who lived there. I do not believe it because this planet would have been too far from the sun. Mars is already too far. I think that a collision with something very big shattered this planet into billions of big and small chunks. Because these chunks were hot liquid right after the catastrophe, they formed into ball and potato-shaped asteroids before cooling to be a solid body in space. Most asteroids remained in the former orbit of that planet, but many went into odd orbits. And still more were ejected out of this orbit, disappearing into space or into the sun.

I am not alone with this belief. The Sumerians – the world's oldest known civilization, dating back almost 6000 years – have bequeathed to us in their writings on clay tablets and in their art extensive and detailed chronicles. These writings mention intelligent beings that came down from the sky, taught them many skills and an amazing knowledge of cosmic events that triggered the formation of today's solar system. And one of these events is exactly the same story about the destroyed planet I just explained. In addition, the Sumerians wrote that one of the very big chunks was captured by our earth and became the moon.

So now we have the asteroid belt between Mars and Jupiter, and thousands of them have such an odd orbit that they cross the path of the Earth again and again. From time to time one hits the Earth. Jupiter can catch only a few of them, and Mars is too small to do a decent job. We are facing a permanent danger that some of them will hit us, as has already happened many times in the history of the earth. Of course, the four giants out there are always doing their job in catching objects that come in from outer space. In July 1994, we had the opportunity to watch Jupiter "at work", when one of the many comets was eliminated piece by piece.

Our earth collects meteoroids all the time, day after day. This rain of small pieces is normal, and our atmosphere takes care of them. But once in a while, about every 40 to 60 million years, we are hit by a very large one, which causes a disaster. Some produce only a very large crater (like the one in Arizona), and the surrounding areas on that continent has big losses of life. But a few are so big that they punch a hole into the solid crust of the earth, resulting in a large lava field in addition to a horrible explosion. These hits become global catastrophes, terminating most of the life on earth. The last of these "big ones" occurred about 65 million years ago, another one about 220 million years ago. And here I have to do some reasoning in support of Chapter IV.

How does evolution redo most life forms every 40 to 60 million years (which needed over 2 billion years to evolve, as the theory goes)? These 2 billion years were not available for continuous evolution for most of all species. There is much proof for this in the geological layers of the earth. The fossils in these layers tell us that after each one of these catastrophes life came back immediately with fully developed plants and animals. And often, when the disaster caused a total change in climate, a completely different line of animals appeared, fully developed. For example: 60 million years ago the era of the gigantic reptiles (dinosaurs) was over, and perfectly designed mammals took over – immediately!

This is bad news for the theory of evolution. Evolution would have a very hard time working that fast. Everything points to the only logical explanation: After each of these disasters the earth was completely reseeded with life forms, matching the new ecological state of the earth.

After the big disaster about 220 million years ago, during the Triassic age, the earth had a super-tropical climate. Vegetation overgrew most of the planet. Almost the whole planet was covered by dense tropical forest. Each time a geological upheaval of great magnitude occurred, large masses of this vegetation were washed into valleys by floods and piled up. Later, they were buried under layers of soil and rock and became the raw material for coal. Terrible earthquakes, ranging beyond a 10 on the Richter scale, were the catalyst for these changes. It was mountain building time!

At that time the earth was even less stable than today. But the vegetation remained out of control because of the supertropical climate. So what was there to do? The entities who are responsible for the development of our planet knew the solution: At the end of the Triassic age, huge plant-eating machines were brought in – the dinosaurs. They inhabited the entire earth and chewed all the way through the Jurassic and Cretaceous ages, some 130 million years in all, until the beginning of the Paleocene age, about 65 million years ago. The vegetation was growing like mad mainly because of the very high carbon dioxide content of the atmosphere at that time. As a result of this condition all plants contained only half as much nitrogen as our plants have today. Because all animals require nitrogen to built proteins, the poor dinosaurs had to eat twice as much, and most plants were very hard to digest. The solution was a huge animal with an enormous internal food processor in the belly – actually a walking fermentation tank!

About 65 million years ago a very large asteroid hit the earth at the North-West Coast of the Yucatan peninsula (Mexico). Even today a crater ring of about 300 miles in diameter is still detectable. (We hear and read about this event all the time in scientific journals and in the media.) The resulting global disaster killed most of all land animals, especially the dinosaurs. But it was not the end of the dinosaurs, many of them survived, mainly the smaller and carnivorous dinosaurs.

What really happened, and what was the actual catastrophe, was the resulting climatic change. The carbon dioxide driven hot temperatures changed to conditions similar to what we have today. Vegetation became normal, and food was in short supply for the giant plant destroyers, the huge dinosaurs. Gradually the dinosaurs became extinct. As the dinosaurs became slowly extinct, mammals were introduced and took over. The huge saber-toothed tiger was one of the first mammals introduced. His job was to eliminate the remaining dinosaurs. When this was accomplished, he himself became extinct after several million years.

Dinosaur fossils are being found in geological layers of the Paleocene epoch and some even in the Eocene epoch, forty-to-fifty million years ago. This proves that they died out gradually over a period of about 10 million years, and not within a few days or weeks after the huge meteor hit 65 million years ago, as so many theoretical scientists believe. I sometimes wonder why scientists do not work together better. Some are at their desk and theorize a big meteor doing it all at once, while others work in the field and find evidence that this could never be. Does not anyone read the papers of the others?

When the Tertiary period ended with the end of the Pliocene epoch, 4 million years ago, the earth was finally ready for the introduction of human beings, suitable for the incarnation of human souls. (Yes, there are other than human souls in the universe!) The first bodies were of a robust, strong, and rough nature, which complemented the rough conditions of that time, but the brain was already large enough for the requirements of a human entity. (The so-called missing link never existed; there was nothing to be "linked".) Life and survival was not easy, but the fact that these people made it through the next 3 million years is proof that the design was good for those times.

Then the modern human was introduced, the Homo sapiens. Readings by Edgar Cayce stated that five different basic races were simultaneously brought into the earth plane, each one on another continent. The black race was placed in Africa, the red race in Atlantis, the brown race in Lemuria, the yellow race in Asia, and the white race in Europe.

Each race came from another planet and is designed for a

certain climate and environmental condition, this being the only difference between them! For example, the black race is designed perfectly for the tropical forests and savannas. The high pigment content of the skin protects against UV-rays; the curly hair is excellent heat protection; and leg length in ratio to the body, together with specialized muscles, provide for fast running. All in all it is the perfect body for fighting the rough conditions of Africa. The white race is perfect for cool and cold climates. Brown is best suited for the humid tropical, etc. (Today these races and their mutations and mixtures make it extra hard for many people to graduate from this school, the earth, because they think that their own race is superior to all the others, which is nonsense. All are equal.) I am sure that there must be many planets in the universe inhabited by only one human race.

The Homo erectus type humans, who made it through the first 3 million years, could not survive against the competition of the more ingenious Homo sapiens who were just perfect – *smart and tough* – for the conditions at hand. A sixth race, the Homo neanderthalensis, was introduced about 70,000 years ago in Southern Europe, but did not make it against the established Homo sapiens (scientists are not sure why). About 25,000 years ago, the last "Neanderthal man" lost against the Homo sapiens.

The above description of the appearance of humans on our planet is only somewhat similar to the stories of our anthropologists. It is about all they have discovered so far, and we better not rock the boat! I say this because there are so many findings that point to much earlier appearances of humans and great civilizations, yet all are entirely disregarded by the anthropologists, because they do not fit into their sacred picture. Each one of these discoveries is ignored and swept under the rug. So just for the fun of it, let us rock the boat as hard as possible.

First, here are a few findings that raise not only questions, but also many eyebrows: An iron cup found embedded in coal in an Oklahoma coal mine; a metal tube recovered from a 65 million years old chalk bed; a gold chain encased in a lump of Illinois coal; a grooved metal sphere taken from a Precambrian mineral deposit; and a steel nail embedded in sandstone in Scotland. These findings are excellent food for thought and speculation. For me they are proof for the theory that life on earth was seeded by people from other planets over the whole history of this planet. When these people came to the earth to "deliver", they had the same experiences that we have – once in a while they lost some items, or left them behind on purpose.

But whatever was going on these millions of years ago, man arrived on earth much earlier then previously thought, and so did great civilizations. The following are a few little known and well-suppressed examples we are not supposed to know, because they contradict the "official" scientific picture.

## Antediluvian Civilizations

Very impressive are the discoveries made by William Niven, a mineralogist, about forty miles northwest of Mexico City. The whole story is very long, so I will give only a summary of the most important and interesting details. What did he find?

Because he was curious about this region from a geological viewpoint, he began to dig down at that location, like digging a well. After only one foot of topsoil, he worked his way down through nine feet of sand and gravel, mixed with small boulders. This layer is definitely sediment, washed in by big ocean waves, probably a tsunami. The fact that this area is seven thousand feet above sea level, and shielded against the Pacific by mountains that are another five thousand feet higher, is not strange at all. Almost every part of our planet was once upon a time under the sea. But this area was never ocean floor; the sediment was washed in by huge waves. This was a coastal area. We have to keep this fact in mind!

After digging through the first nine feet, he hit a pavement made of concrete. In the nine feet of gravel above this pavement he found fragments of broken pottery, small clay figures, spear and arrow heads, spindle whorls and other artifacts. This tells us that the sand and gravel was not deposited, but that a giant wave must have destroyed a town while washing in the sand and gravel.

Niven broke down through the pavement and confronted another six feet of gravel and sand, this one without any man-made artifacts enclosed. At the bottom he found another pavement, made of bricks and squared rocks. It seemed to be a road.

Below the second pavement was a third layer of gravel and sand, this one fourteen feet thick and with no man-made artifacts. Then came a three-foot thick layer of volcanic ash. Below that, about 35 feet from the surface, he hit an anthropologist's dream: A third pavement of concrete and on it house after house buried by the volcanic ash. The houses are fancy, the walls painted in beautiful patterns and some inlaid with mosaics. Many decorations are made from metal or wood, but the walls are solid mason work. Everything made from wood was petrified to hard stone. The most exciting find was a complete shop of a goldsmith and jeweler which contained hundreds of items: Many tools, over two hundred models for die-casting, and some finished gold ornaments. The furnace still had gold in it. The guy who worked here also made jewelry from Jade, and Jade is not a mineral found in Mexico.

Niven also found small figurines of a Chinese man; this figure was in a lotus position (yoga). Hundreds of skeletons were also found, all looking solid, but they crumbled into dust just by being touched. In one house a room was nicely decorated for a little child.

Some experts dated this city at 70,000 years old. However, that would mean that this area was at sea level and the mountains were raised later, not millions of years ago as geologists argue. Other scientists said that this area was less than a hundred feet above sea level at the end of the

Pleistocene epoch, 200,000 years ago. This would be a better match for the petrified wood items, the clay models that became hard stone, and especially the powdery skeletons. (Edgar Cayce mentioned 220,000-year-old civilizations.) What we also have to think about in respect to the coastal mountain ranges is the fact that the upper pavement indicated two disasters younger than the low city! These mountains then cannot be very old, certainly not million of years! What do scientists say today? Nothing! They successfully swept it under the rug.

Then we have the very big problems, very big in the exact meaning of the word.

About forty miles east of Beirut, Lebanon, is the acropolis of Baalbek, built by the Romans as sanctuaries of Jupiter and Venus. So far, so good but one part of the enclosure wall, called the trilithon, is composed of three blocks of hewn stone, each weighing almost 800 tons. They were lifted 22 feet in order to lie on top of smaller blocks. They came from a quarry downhill from that location. There lies the fourth block, nicely finished to a size of thirteen by thirteen by seventy feet, weighing approximately a thousand tons. The Romans did really great things, but these blocks are far above their capabilities. Who did it? How did they cut such giant blocks out of the rock in the quarry? How did they move them uphill? And how did they lift them 22 feet? If we consider our most powerful crane, a gantry crane at the Grand Coulee Dam, which was tested at 2,500 tons, and runs on rails inside a very strong building, then we can be very sure that no company would take on such a job because the crane had to be a monster-size steel construction from the quarry to the wall. But how about portable cranes? With their limit of 400 tons we can forget about it.

The 228 feet high Black Pagoda in India is capped with a single stone slab that is estimated at over 1,000 tons, the equivalent of five heavy locomotives. How did they get it up to the height of about 220 feet? And who did it without modern machines like ours? (It is not even established that we have machines that could do it.)

And of course, there are the pyramids of Egypt, especially the so-called Great Pyramid, the largest of three. This work also exceeded the capabilities of the people who supposedly built it. I have already mentioned that it is very hard to believe that the Egyptians could quarry, transport, and place into position 2½ million blocks, each one weighing 2½ tons, every 1.7 minutes. Cheops had only twenty years available! (I also explained the great forgery of 1837 which asserts Cheops did not do it.) For the roofs of the chambers and the great gallery, dozens of much heavier blocks had to be moved and put precisely into place. Most of them are around 400 tons each.

And then there are questions about the design. The mathematical factor *pi*, the ratio of the circumference of a circle to its diameter, is embodied in the ratio of the pyramid's height to the perimeter of its base to a precision of five decimal points, 3.14159. Our engineers do not have to use better than 3.14 for most of their calculations. The perimeter at the base equals a half-minute of equatorial longitude, or 1/43,200 of the earth's circumference, and the pyramid's height equals 1/43,200 of the earth's polar radius. This shows that the designer of the Great Pyramid had very good knowledge of the shape and size of the earth!

This pyramid was not built to be a tomb. Khufu (Cheops) had his grave among the other graves of the pharaohs. The tombs of most pharaohs are covered with hieroglyphs and cartouches; in the Great Pyramid there are no inscriptions, none at all! What is really interesting is that, according to some inscriptions from the first dynasties, the Egyptians believed that their civilization had a much greater antiquity than our present experts believe, one that reached more than thirty millennia into their dim past.

And now let us go around the Globe to South America. There, at the southern border of Peru, is Lake Titicaca, 12,500 feet above sea level. At its shore we find the ruins of the ancient city of Tiahuanaco. Irrefutable evidences are present in and around this ancient city, which prove that when it was built the ground on which it stands was just above sea level. It was a harbor city. This city is estimated to be about 50,000 years old. The massive stone blocks of the harbor walls were joined with heavy T-shaped silver bars. (The Spaniards removed them.) The age of this harbor city tells us that the Andes cannot be millions of years old; they were lifted up less than 50,000 years ago. And if we look at any picture from the Andes, everything is very rough, steep, and not smoothed by the weather – in effect, very new. The Andes are a young formation!

These few examples show clearly that we had great civilizations on earth far into the distant past, at least as far back as 200,000 years. Terrible cataclysm erased most traces of these societies, and most of them are now at the bottom of the oceans.

Earlier, I mentioned that the Mayan language is spoken both in Yucatan, Mexico, and by the Naga people in the Himalayas. Also, half the words in Japanese have Mayan roots. How is it possible that this language can be found on both sides of the Pacific? And why are many dialects of the Pacific Islands similar to Maya? Why do we find Maya symbols on ruins on some of the larger Pacific islands? On the Rarotonga Island is a paved road coming out of the ocean, crossing the island, and then disappearing into the water on the opposite side. It is a very small, uninhabited island, but this road sends a big message: There was a continent in the Pacific that went down in the very distant past, and the language of the people must have been Mayan! The Mayan speaking areas of today are the remnants of former colonies of this sunken continent. Many carefully researched books about this continent Lemuria (or Mu) are being written. And the traditions of Lemuria point to another happening: When this continent went down in a catastrophic disaster overnight, something else had to go up – the Andes and the mountains of Central America.

On the other side of the earth is the Atlantic Ocean, which got its name from the sunken continent of Atlantis. There are many books about Atlantis on the shelves of our bookstores, three of them are already classics: *Atlantis, the Antediluvian World* by Ignatius Donnelly; *Edgar Cayce on Atlantis*, by the ARE in Virginia Beach; and the writings of Plato, the Greek philosopher (427–347 B.C.), who learned about Atlantis from Egyptian writers. According to Edgar Cayce and other sources, the civilization of Atlantis had such nice toys as airplanes, power-driven ships, and submarines during their last centuries. They destroyed their continent through misuse of their technology (at 10,500 B.C.). Plato's writings also suggest this.

I have some delicious food for thought. There is a tribe near Timbuktu in Africa called the Dogon. They know something that our scientists learned of just a few decades ago. For many centuries they have a very important annual festival in honor of the little companion star of Sirius, the Dog Star. But this small companion star, Sirius B, cannot be seen with the bare eye. It is absolutely impossible! Even with a big and powerful modern telescope it is very tricky to separate it from the gigantic Sirius A. Sirius A is so extremely bright that it blinds out the little one, which is very close to Sirius A. How can the people know about the existence of the small Sirius B? And why do they believe that their ancestors came from there, and celebrate this tradition annually? I think it is interesting and something to think about.

## The Geology

It is a generally accepted theory by our scientists that the earth originated out of a ring of cosmic dust and gases around the sun. We also can be sure that a high percentage of these metal and non-metal particles were not pure elements, but compounds of two or more elements. Many of these elements were gases, especially oxygen, which happily combines with almost every other element, especially hydrogen. (The best example is water, $H_2O$.) When the mass of the ring was finally condensed into a huge ball, it collapsed under the force of gravity and became hot liquid. Then the surface cooled and developed a solid crust. Inside this new earth were not only big pockets of trapped gases from the ring, but even greater amounts of gases were released from the chemical compounds by the heat.

And this process is still going on today. Because much of the earth is a viscous magma (molten, liquid rock), these gases work their way very slowly up to the surface. While on their way up, smaller gas pockets unite with others, and finally we have huge gas chambers right under the solid crust.

Before I continue, a word about natural oil and gas deposits. I just cannot believe in the widely accepted theory that oil and natural gas formed in the age of the dinosaurs, when lush vegetation was buried and subjected to high pressure and heat. (I mentioned previously that such a process produces coal.) Even though all textbooks say so, I believe otherwise. I think oil and natural gas came with the planet from the very beginning; they must be primordial. Oil and gas are just two of the many chemical compounds of the original earth.

This oil oozes very slowly from the original deep deposits towards the surface of the earth and becomes concentrated in so-called oil fields, where it can be reached by drilling. And why are oil wells so rich in helium? Helium was never a chemical component of plants; so it must have been absorbed by the oil on its way up. Logic tells us that helium must be trapped in high amounts inside our planet because it fills the space of the universe, and therefore was part of the primordial particle ring around the sun.

I shall return now to the big gas chambers under the solid crust of the earth. The gases that move up from the inner earth are under extremely high pressure. As they enter the chambers below the crust, the pressure constantly increases. Gases can be compressed and therefore nothing happens for a while.

Huge gas chambers can even hold up a whole small continent.

But eventually the pressure becomes too much for the relative thin crust of the earth. First, cracks in the lower region of the crust result in earthquakes. In the next stage the crust lifts, producing high altitude areas and mountains. Finally, the moment arrives when the crust cannot hold the pressure anymore. The pressure forces a way through the solid crust, and a volcano is born. In this case an explosive-type volcano, ejects hot gases and ash high into the atmosphere. (In contrast to the volcanoes caused by fault lines, which eject lava.) The volcanoes provide a fragile balance for a while, a "while" being millions of years, but at the end the story is always the same. The pressure gets much too high; the crust bursts open, and all the gases escape in an explosive catastrophe. Then the crust breaks down into the now empty cavity, and we have an ocean where there once was land. Lemuria in the Pacific and Atlantis in the Atlantic went down in such a way.

When the earthquakes of today are caused by gas pressure from below, and not from faults, the surface of the earth is always lifted a few inches or even feet. These earthquakes normally have thousands of aftershocks, because their bulges are still cracking in thousands of spots. Earthquakes caused by faults do the whole job more or less in one strike, and we have longer intervals between fewer aftershocks (in many cases years). But gas underneath keeps pushing until the whole thing stabilizes, permanently cracking (aftershocks) during this process. It is my opinion that the Los Angeles earthquake of January 1994 was of this kind, because the ground was lifted, and that the so-called Whittier Fault is not a fault, but rather a gas chamber deep in the earth.

In the spring of 1994 we heard about an earthquake measuring 6 on the Richter scale, centered four hundred miles deep under Bolivia. Then there was absolute silence about it. Why? I guess nobody had any idea how to explain an earthquake within magma, which is liquid, molten rock. There cannot be any faults in this soft stuff! Soft? Liquid cannot be compressed and movements are transported immediately through it. May I suggest that two already big gas pockets were joined happily into one even larger one?

Many areas on our planet are still rising slowly, steadily, and smoothly. This is true for most of the Andes, New Zealand, and even some flatland areas. For example, the Great Lakes area is slowly tilting and will eventually empty to the South. There are also many areas of the ocean floors which are on the rise. Some rise slowly, but a few suddenly rise a hundred feet or more, causing a tsunami (wrongly called a tidal wave).

For all these rising areas the theory of gas chambers makes much more sense than faults. Some geologists will say that such chambers should be detectable by sonic testing. That is right, but we also have to consider that these gas pockets are under such tremendous pressure that the gas is liquefied! So the technician will read "oil" or "magma" instead of liquefied gas.

The type of volcanoes in a certain area seems to be a good hint of what is underneath. Lava volcanoes, as in Hawaii, Iceland, or the Aetna, are typical for fault lines. Exploding volcanoes that put out cubic miles of gas and ash are a sign of gas chambers underneath, like the "ring of fire" around the Pacific, and the volcanoes of the island chain of the Antilles.

Everyone has this question: "Where on earth are we safe?" I would like to help you to find the answer by yourself. Just read and think about this paragraph. The average diameter of a chicken egg is 43 millimeters. The thickness of the shell averages 0.4 millimeter. Therefore the ratio of shell thickness to the diameter of an egg is 1:108. Our planet has a diameter of 12,757 kilometers. The solid crust averages 60 kilometers (38 miles) thick. Therefore the ratio of the thickness of the solid crust to the diameter of the earth is 1:213. This is twice the ratio as for the egg. To get the right picture about the earth, we have to imagine an egg, in which the shell is only half as thick as normal! I do not believe that this makes you feel very good. With this flimsy egg in mind, now we can easily answer the question of "where are we safe?" *Nowhere*! The moon makes this picture even worse. In addition to the tides of the oceans, the moon causes a tide of the earth's crust. Not much, only a few inches, but because this happens twice a day, it is enough to produce the initial cracks for earthquakes over time. I hope that this picture of the extremely thin crust of our planet explains why there are permanent earthquakes and volcano eruptions day after day.

One very popular story we have these days is that global warming by the greenhouse effect will melt the polar ice caps; as a result, the sea level will rise. But there are a few scientists who see an opposite effect. They explain it in this way:

The somewhat warmer water will melt more ice at the edges of the ice caps, but the polar regions always had, and still have, an extremely dry climate with very little precipitation. (The Antarctic has only two inches of rain equivalent per year.) With global warming, more water will evaporate and more clouds will reach the poles, resulting in more snow which will remain there. So, theoretically, the sea level will fall. But very likely more ice will melt at the edges of the pole caps, and most glaciers will recede. This will then just compensate for the extra water that remains in the pole caps. The just perfect size of the pole caps is the reason for this compensation. I have already explained that the famous 23,5° tilt of the rotation axis of our earth is the reason for this stabilizing effect of the polar regions. It has also to do with the self-healing capacity of our planet.

It is safe to say that the earth is still far away from being stable. The crust is not adequately thick and there is still too much gas inside. And in the surrounding space of the solar system are too many asteroids in orbits too close to earth, ready to cause many great disasters in our future.

In effect, it is too early for the earth to harbor a peaceful, global super-civilization, lasting for thousands of years, or even a few hundreds of thousands of years. Such a civilization would have to have the weather, the asteroids, and the comets under full control. The earth's present population will never be able to develop such a perfect civilization. If the people on this planet cannot live in peace and harmony at home, how could they have peace worldwide? Of course, many dream about a world government and worldwide peace, but let us face it; it is really only a dream. What would they do with all the millions of criminals? There seems to be an endless supply of new criminals. No, we better forget about the idea of becoming a super-civilization.

## Insights

Now that we know how unsafe the earth is, we can understand why this planet was chosen to be a special planet for the rehabilitation of the "bad guys" among the souls – those who tried to do their own thing in violation of the laws of the cosmos. For this purpose is the earth not only just good enough, but perfect.

Maybe after another few million years, this planet will be ready for incarnations of free souls without bad karma, and a super-civilization can begin. Maybe those souls will be incarnated into better bodies with a better brain, if such is possible or necessary. I doubt that a better brain will be required for two reasons. First, a healthy human body, beautiful and possessing intelligence, is pretty close to being perfect. And the brain does not have to be larger. It only has to have an open connection with the ethereal level. Only then will we have a spiritual-oriented civilization, in harmony with God and reaching the highest possible levels in science and technology. (To be sure, being spiritual-oriented is not the same as being religious or being dictated by church dogmas, which is believing without knowing and therefore wrong.) The second reason is a physical one. The opening in the pelvis of a woman is now just large enough for the passage of a child's head, in many cases already too narrow. If the heads would get bigger, we cannot be born in a natural way, so we will have to do with what we have!

Right now we have many good souls incarnated here on earth, which are through or almost through with their karma, and do not have to reincarnate into another human experience. But the problem is that there is always a steady stream of bad newcomers who spoil what the good people have accomplished. The result is the temptation to give up, just because of this seemingly hopeless condition. But we should never give in to this temptation if we want to get out of here!

Finally I would like to present a different look at our planet, from an entirely different viewpoint. (It is not entirely my own theory, but a compilation of information from dozens of books and even videos.) In Chapter II I hinted that the belt of the constellation Orion seems to have an important meaning. There is a conspicuous similarity between the arrangement of the belt-stars of Orion, and the three pyramids of Giza. These three pyramids are oriented to each other exactly as the three belt-stars of Orion. The third pyramid, the smallest, is off the straight line of the two big pyramids in exactly the same ratio as the third star of the Orion belt. And there is more: The three pyramids are oriented in ratio and position to the Nile river in the same way as the Orion stars to the Milky Way. Why? It must be so on purpose.

There are some interesting connections. The brightest star in the sky, Sirius, which I mentioned earlier in connection to the Dogon tribe, is located in almost the same general direction as Orion. This means Orion could be a perfect signpost for directions to get there. In the same general direction, but on the other side of Orion, we find the Pleiades. In order to go there Orion would be again the best signpost. Both cases work fine because Sirius and the Pleiades are relatively close to us, but the belt-stars of Orion are far away.

If we consider the fact that many groups of people here on earth have traditions about other stars, and about ancestors who came from there, this "coincidence" gets interesting. There is the Dogon tribe in Africa which has the tradition about the small companion star Sirius B. The Mayas had their calendar of important days oriented to the day in January, when the Pleiades are overhead. And the Mayas and the Sumerians left writings behind that point to ancestors in the Pleiades.

According to the literature that deals with these interesting connections, the three pyramids of Giza are not built only for the purpose of initiating people, but serve another important purpose. They are a sign for travelers from the Sirius system, or from the Pleiades, that they have arrived at the planet where their descendants live. Some day we may have proof.

If we consider the hundreds of traditions and written records about such ancestors or teachers from outer space that we have on earth, we have to accept the idea that we are connected to the universe in many unexpected ways.

# Chapter IX
# A NEW THEORY OF THE UNIVERSE

> To merge and to separate, to create and annihilate, this is the way of nature's eternal change.
>
> The Yin and Yang principle of the Tao

First let us take a look at the picture of the universe, as we receive it from our astronomers and astrophysicists today. As in previous chapters, I will first mention current theories and then mention other interpretations, or what I think about it.

Before the universe existed, there was absolutely nothing; no space, no time, no matter, no energies. The stuff of the future universe was squeezed into a mathematical singularity of infinite tiny dimensions, the so-called cosmic egg. Then, about 12 to 15 billion years ago came the Big Bang. This singular point of nothing exploded, and matter, time, space, and all energies rushed out into the nothing. At the beginning of the Big Bang there was nothing, but less than a second later the entire universe had sprang into full being.

Of course, all of this contradicts any kind of physics or logic, but it is the currently accepted theory about the beginning of the universe. Astrophysicists all over the world are very busy developing sub-theories and sophisticated formulas in order to find out what happened during each millionth of a second during the first milliseconds of the Big Bang.

The theory also goes that the objects that are farthest away, about 12 billion light years, are believed to represent the universe as it was about 12 billion years ago, shortly after the Big Bang. (The light from these objects needed 12 billion years to reach us.) Behind these farthest galaxies are even more distant objects, the quasars, but their real distance can only be estimated – some say 12 billion light years, others believe 20 billion light years. Because this idea conflicts with the Big Bang idea as an explosion, a reader of the *Astronomy* magazine had this question: If the Big Bang was a big explosion, and everything, space included, expanded into the nothing, then how can we believe that the most distant objects we see were the beginning? The editors, after consulting an astronomer, answered that the Big Bang did not happen in some central place, or out there; it happened everywhere, here included. Okay, if we need a good example for the term "contradiction", here it is. An infinitely small singularity exploded all over the universe simultaneously – beautiful!

If we accept the basic idea of the Big Bang, and apply just common sense to it, then we could see it this way: If the universe expanded after the Big Bang at half the speed of light, which is a very generous assumption and very unlikely, then these objects needed 40 billion years to get to the most distant location of 20 billion light years. From there the light needed 20 billion years to reach us. That makes at least 60 billion years for the age of the universe, if the Big Bang in fact occurred, which I do not believe. We see a very twisted logic behind the Big Bang theory. But it gets even better.

First, if a sphere-shaped object explodes, the debris speeds away into all directions. A while after the explosion the result is a hollow sphere, made up of pieces that fly away, and this sphere is empty inside. So what are we doing here in the center of the universe together with all the other galaxies? The universe should only consist of an outer sphere-shaped shell. That is the right picture after an explosion, and not with the highest density in the center, as we have in our universe.

Second, everything moves away from us (so we must be in the center). The closer objects move at a lower speed, and the farther they move away from us the faster they go. This observation contradicts the principles of physics. The debris of an explosion goes fast initially and gradually slows down.

Measurements show that the most distant objects speed away with almost the speed of light. Because the physics for explosions does not allow for acceleration, the initial speed of all objects right after the Big Bang must have exceeded the speed of light by magnitudes, which is impossible according to Einstein.

The latest observations (and calculations) show that the universe is only about 10 to 12 billion years old, while many stars are measured to be 16 billion years old. This paradox of the children being older than the parents is a huge headache right now.

Another problem is the fact that the most distant galaxies are elliptical, while in our region most are spiral. Theories about how galaxies form requires spirals for the beginning and ellipticals for the end – but the ellipticals out there are supposed to be at least 12 billion years younger than spirals like us.

The Hubble Space Telescope provided us with high quality pictures of objects in space, which answered many questions. But because of the tremendous power of the Hubble, it also introduced more questions than it solved. One of these is the fact that the most distant galaxies are not only very odd in shape, but also very small, some only 2,000 light years across. It would take hundreds of them to combine into the size of the Milky Way galaxy, and it is absolutely unthinkable that they can grow to regular size in the normal way of collecting matter from its surroundings. Something is basically wrong with using telescopes for looking back to the beginning of the universe. (Looking back in time is, of course, what we do by using them, but not back to the beginning!)

And certain things are even funny, at least for non-scientists. For example, the famous black holes are believed to grow as big as the mass of a whole galaxy. Nothing gets out, not even light. What is the fate of a black hole? For many years it was believed that black holes will exist forever because by the laws of physics they cannot explode. Then came Stephen

Hawking, the English physicist, with the solution for the death of a black hole. Through quantum-tunneling, or quantum-evaporation, a black hole will evaporate extremely slow. It is called "Hawking Evaporation". How long will it need for one of the large black holes to disappear? At least $10^{92}$ years! That is a 10 with 92 zeros, really a long time! How long could that be?

When I was a little boy, I asked my grandfather how long eternity is. He explained it to me this way: There is a big mountain, so high that it always has snow on top. Every 100 years a small bird arrives and grinds his beak on the rock of this mountain, scraping off a few particles of the rock. When in this way the whole mountain is gone, the first second of an eternity has past.

In my opinion, the only way to arrive at such exotic ideas is to develop fancy mathematics first, and then try to explain the result with a theory. Good science should be the other way around. First the idea and theory, and then mathematics or test, just the way they did it in ancient Greece. (And Einstein was a strict believer in this method, too.)

I believe that the whole Big Bang idea is just one enormous paradox, contradicting itself in many respects. But some contemporary scientists believe to have proof for the validity of the Big Bang theory, namely the results of the COBE project. COBE, the Cosmic Background Explorer satellite (1989), measured the so-called cosmic background radiation in the microwave range. The Big Bang people believe that this microwave background is an echo of the Big Bang. The pattern they received from COBE is their proof for the validity of the mathematical picture of the Big Bang. The problem is, that there are as many scientists who have come up with many different possibilities for this microwave background, that have nothing to do with the idea of a Big Bang.

Why do we have the theory of a Big Bang anyway? The reason is the work of the American astronomer Edwin Hubble, about 70 years ago. When he checked the red shift (displacement of spectral lines by the Doppler effect) of other galaxies, he saw that all galaxies in the universe are moving away from us. And because the galaxies move faster at greater distance, he developed the so-called Hubble constant, a factor to find out how much faster a galaxy moves away, based on the measured red shift. (The farther away, the stronger the red shift.)

Then the theoretical scientists tried to find out *why* it is that way. Many scientists tried their luck, but the winner was the cosmologist George Gamow, who suggested the idea of a hot Big Bang. (Of course, it never explained the acceleration problem for distant objects.) Almost everybody adopted this theory and they developed the strange picture which I explained so far.

## A New Approach

Now I will present my own theory of the nature of the universe, based on common sense and logic. Please remain open-minded. Even if it requires imagination that is stretched to the limit, it can easily be comprehended.

Let us first talk about infinity. This is a hot iron that most scientists refuse to touch with a ten-foot pole. Why? There is no reason to stay away from infinity. If we approach infinity with an open mind and some logic, we will find a few comforting facts.

First we must accept that our brain is designed for living in a physical world. All we can comprehend are three-dimensional conditions and, after some learning, even something four-dimensional. Three dimensions means that something is a certain length, width, and height. Then, if we add the time that an object is at a certain location, an airplane for example, we have four dimensions. But we can perceive only these dimensions; everything has a limit. We cannot grasp anything without an end. But, our brain is capable of logical reasoning! And logic tells us that space and time must be infinite! So let us see it this way, and let us accept the fact of infinity, even if our brain does not allow us to comprehend it.

Since infinity is the only factor in all of my theories that cannot be comprehended, I have just explained the reason for this fact. So let us go to the basic foundation of my theory.

I distinguish between *cosmos* and *universe* in my theory. The cosmos is everything there is and infinite in size. This infinite cosmos contains an infinite number of finite universes. (I use these names because they are handy and I did not want to invent new names.) This infinite cosmos with an infinite number of universes has existed since eternity, and will exist forever; time is infinite too.

And now I will apply some logic and common sense to the idea that our finite universe is just one among an infinite number of universes. The earth rotates. The moon orbits around the earth. The earth orbits around a rotating sun. The whole rotating solar system rotates around the center of a rotating galaxy. Even in the subatomic world everything rotates. Everything we observe rotates or orbits; so *why* should the universe be the only exception and act like a firecracker? I think it is much more likely and logical that all universes rotate like everything else, including ours. Rotating and orbiting are the basic requirements to keep everything going and in balance!

I talked about Edwin Hubble and the red shift problem. The term red shift appears again and again in our publications, but I think it should be explained for the non-scientist. When an airplane approaches us, the pitch of the engine noise is high, because the relative motion between the source and the observer shortens the wavelength of the sound. It is called the Doppler effect. Then, when the airplane flies away, it has the opposite effect, the wavelength is stretched and the tone is low. Exactly the same happens to light, when the source is moving away or approaching. The wavelength of the light is changed to longer wavelengths, towards red, when an object rushes away. This can be observed in the location of the dark lines in the spectrum of a star. If an object moves towards us, the wavelength shortens (to blue) and we talk about blue shift.

And this red shift is the mother of the Big Bang theory, because it is the only explanation for the expansion of the universe our scientists can think of. But my theory explains not only the expansion in another way, but also accounts for the acceleration problem. Let us go step-by-step and investigate the other possibilities in my theory, which seem to me much more likely.

Currently there is a hot debate among astronomers about the so-called great attractor in the universe, which lies somewhere in the direction of the constellation Centaurus. This theory was proposed because an immense stream of galaxies, including our Milky Way, are moving like a river of galaxies, all at the same speed and in the same direction. But the problem is that this river flows in the wrong direction, perpendicular to the direction of the expansion of the universe. Everybody believes in a "Great Attractor" who causes this river of galaxies to flow, but so far this object has not been identified. (There is even a whole book about the search for the Great Attractor.)

The head of the Mt. Stromlo observatory in Australia, Donald Mathewson, was skeptical of this theory and the conjectures of others. He measured the data from five times more galaxies than the others, and did so with extreme care and accuracy. As a result the group was able to nail down the exact speed of this stream – about 450 kilometers per second.

What does that mean? First, it eliminates the idea of a great attractor, because the speed would then vary. Gravitation gets weaker by the square of the distance. Therefore the first ones, closest to the attractor, should move much faster than the last ones. But why do they move perpendicular to the direction of the expansion of the universe?

The stars in our spiral arm of our galaxy seem to move together in one direction, all at the same speed and perpendicular to the radius of the galaxy. But we know that this is not caused by the gravity of some attractor. We know that it only looks like that, because we all rotate around the center of the galaxy and, because we are all at the same distance from the center, at the same speed.

The exact same condition applies to the problem of the great attractor. This river of galaxies rotates around the center of the universe, and that is why we all have the same speed. And if we should find a huge galaxy cluster ahead of us, it is not the imaginary attractor, but just another group that rotates ahead of us at the same distance from the center of the universe.

Now we will find out where we are within the universe, and what are the conditions. To do this; we have to employ a different kind of reasoning. Our sun rotates around the center of our galaxy at a speed of 220 kilometers per second. One ride around the galaxy requires 230 million years, and our distance from the center is about 32,000 light years. If we look for a planet in our solar system that has about the same ratio of distance from the sun (as the sun from the center of the galaxy), it would be Saturn. Saturn orbits around the sun at 9.6 kilometers per second, the sun rotates at 220 kilometers per second, and the river of galaxies we talked about, moves at 450 kilometers per second. This shows that the rotating or orbital speed goes up with the size of the system. Even though a "small" system like our galaxy spins faster than a universe, the fact that 450 kilometers per second is only about twice the speed of the sun suggests that we must be relatively close to the center of the universe. It looks like the huge supercluster Virgo, which is about 30 million light years away from us, is in or near the center of the universe. Virgo is very big and contains thousands of galaxies.

Calculations of the red shift resulted in expansion speeds of close to the speed of light for the farthest objects, about 12 to 15 billion light years away. Some estimate that they will reach the speed of light at a distance of approximately 20 billion light years. Why and how is it that the speed of expansion goes up the farther out the objects are? It is contrary to the physics of an explosion, where things slow down. Okay, we have the

Hubble constant, currently agreed on as 75 kilometers per second per megaparsec (one megaparsec equals 3.26 million light years). As the distance increases, so does the expansion rate. But the Hubble constant is only a mathematical tool. It tells us that all objects accelerate and how much, but does not address the question why and how this happens. Fragments of an explosion do not accelerate; they slow down! I wonder how anybody could develop the Big Bang theory, based on these findings and formulas by Edwin Hubble. It contains no logic whatsoever.

Now I introduce my explanation for the expansion *and* acceleration in the universe. Let us assume that the rotating speed of the universe goes up in a linear function (continuously without change in speed) in relation to the distance from the center. This means that the whole universe rotates like a huge wheel, just like the galaxies do. But such condition would mean that centrifugal forces would be at work on everything in the universe. And the farther out, the stronger it will be, because the velocity enters the multiplier of the equation for centrifugal force squared! (Centrifugal force equals mass multiplied by velocity squared, then divided by the radius.) Also, rotating like a wheel means that objects are not orbiting. Orbiting around what? There is no sufficient mass anywhere in the universe to provide the required gravitation for orbiting.

So what will the rotating speed be, if the speed goes up in a linear way, as in a wheel? Let us take the accepted distance from Virgo, 30 million light years, as our distance from the center of the universe. If we divide the currently estimated maximum radius of the universe, 20 billion light years by our distance from the center, 30 million light years, we arrive at a ratio of 1:666. If we multiply the speed of our galaxy by this ratio, we get $450 \times 666 = 299{,}700$ kilometers per second, almost the speed of light (300,000 kilometers per second). This sounds to be too good to be true, but why not? If this is true, then it means that the objects at the outer rim reach the speed of light and cannot remain in the universe, because further acceleration would mean exceeding the speed of light, and that is impossible. Why is it impossible will be explained in the next chapter, towards the end of my new theory about the nature of gravity. Here I will only say that the reason for the impossibility is not what Einstein thought it is (mass becoming infinitely large).

As a result of these conditions, the size of each universe is limited by the speed of light to approximately 40 billion light years in diameter. Because the rotation speed at the rim is the speed of light, "c", this speed enters the multiplier of the equation for the centrifugal force *squared*! Because of this, $c^2$, the expanding speed will also reach the speed of light; the object has to leave the universe. It will escape into the vast "inter-universal" space of the cosmos, beginning a long journey until it enters another universe. But I will cover that story later in this chapter.

Before I continue with my theory about the nature of the universe, a few things about the speed of light should be discussed. It began in the nineteenth century. After many experiments, some very clever, established the surprising fact that light has always, under all conditions, and within all possible reference frames, the same speed, the world of science did not know what to do with it. This behavior of light contradicted all laws of physics, especially the concept of relative motion (Galileo). Albert Einstein went to work to solve this problem. In 1905 he introduced his solution under the title "Special Theory of Relativity". By declaring indirectly that common sense seems to be wrong, he developed a complicated theory, explaining his solution. At the end he came out with the bold statement that energy equals mass, and that each can be transformed into each other. The equation was the famous $E = mc^2$, which proved to be right in many ways, the most visible indirect result being the atom bomb.

But the rest of the theory about the behavior of light has strange consequences. First, light can be seen as a waveform or as a stream of particles (photons), whatever is most convenient for the situation at hand. (I do not like this double personality.)

Second, at the moment an object reaches the speed of light, the object has zero length, and time stands still. In short, the object and time cease to exist. How can that be? If time does not exist anymore, speed is not possible anymore, because time is a mandatory part of speed, kilometer per second for example. Therefore the above cannot happen. It is a perfect paradox.

Third, nothing can be accelerated above the speed of light, because any object will then have an infinite large mass, and no energy in the world can accelerate it any further. What I see here is that at the moment a mass disappears (see above), this now non-existing mass will become infinitely large. Here no logic is involved, whatsoever.

I hope that nobody blames me for not being comfortable with these three strange "facts" of the Special Relativity. My own theory about light in Chapter X, based on logic and common sense, will make clear why I cannot go along with the above.

And now I will explain the reason for the accelerating escape speeds within the universe, as I see it. It is not a noisy Big Bang. Actually, it is a very simple process. Let us start with an easy to understand example, the earth and the moon. (Everyone with a high school education will have no problems with this.) Billions of years ago the moon orbited the earth much closer. Both, the earth and the moon, collected bodies and particles from space, meteors, comets, asteroids, and other debris. Now we have to realize that the surface of the moon is much larger than the surface of the earth in ratio to the volume. Therefore, the moon gained more on mass through space debris than the earth, relative to the volume. What must happen? In the equation for centrifugal force, mass is multiplied with velocity squared. Therefore, if the mass increases, the centrifugal force will become stronger. Very slowly, over millions of years, the moon will spiral away from the earth, at the moment at a little more than one inch per year.

If we apply the same reasoning to the galaxies of the universe, the picture changes dramatically. While the moon has a velocity of only 0.5 kilometers per second, we are now dealing with 450 kilometers per second (our galaxy), 10,000 kilometers per second, and step by step all the way up to the speed of light, 300,000 kilometers per second. In the equation for the centrifugal force these velocity values enter the multiplier squared! This translates into much more than an inch per year. As a result the galaxies spiral faster and faster to the outside of the universe the farther out they are. First only a few million kilometers per year, but at the end 300,000 kilometers per second!

And that is the real reason for the expanding universe and the red shift. The acceleration is no mystery anymore. (Of course, the radius, the divider of the equation, also gets much larger, but only linear.)

Another factor that enters our formula for the centrifugal force is the mass of an object. In our example of earth and moon, we saw that multiplying the small mass of the moon with a very low velocity squared results in only one inch per year. But on the universal scale it looks entirely different. Just like everywhere in nature, the larger ones eat the small ones. Suns eat planets, gigantic suns swallow smaller ones, the center of each galaxy devours sun after sun, and finally a whole galaxy ends up as one gigantic black hole. Because these huge masses are then multiplied by the square of the above mentioned high speeds of thousands of kilometers per second, now these masses really count in the equation for the centrifugal force.

The final stage for a galaxy to end up as just one huge black hole begins for some galaxies already at a distance of about 10 billion light years from the center of the universe. However, most disappear from our view because they are now only a black hole, about 15 billion light years away. Within the range of 15 to 20 billion light years only black holes are left. We cannot see black holes, and for this reason space appears empty behind the 15 billion light years zone. But this region is as populated as the rest of the universe – with invisible black holes.

What about the galaxies? Because galaxies are rotating like a big wheel in the same way as the universe, the same centrifugal forces are at work to hurl all suns out of the system. But within galaxies we have one very big difference, namely the "very short" spaces between the suns. The universe is very open, the voids are immense, but galaxies are dense in comparison. This condition causes gravity to be the stronger counterforce. Galaxies shrink slowly towards the center. It begins at the center or at two or more centers (Andromeda seems to have two). There the suns get too close and collide, resulting in a star that is just too big. It will collapse under the force of gravitation and we have a so-called black hole. But it is not a hole at all, just the opposite: an extremely dense body in space. The gravitational power is so strong, that nothing can escape, not even light. Because the position of a black hole is a black spot in the sky, it was named in this way, but this name is very misleading. (My two theories of gravity and light in Chapter X will show that is not gravity that keeps light from escaping.)

Once a black hole is born, it begins with its destined job of devouring the galaxy from the inside, solar system after solar system, gaining mass and gravitational power. A black hole can only grow larger, and it can never explode. No known force could overcome so much gravitation. But there is a way to destroy it. How this can be done, will be explained a little later. It will not be by Hawkins Radiation over an eternity, it will be very quick.

While every galaxy is on its way to the rim of the universe, a black hole is busy at the center to consume the galaxies from the inside, until only a black hole is left. Observation shows smaller, less complete, and odd-shaped galaxies, the farther out we look. These strange galaxies are not young as currently believed; they are dying!

Brand new and young galaxies must be extremely large spirals of cosmic dust, with some young stars mixed in. Only the center is somewhat complete. A good example is M100 in the supercluster Virgo, which is located in or near the center of the universe. The galaxies of Virgo should be around 6 to 10 billion years old.

Better developed, but still young, galaxies like ours seem to be about 15 billion years old. We have still many nebulas in the Milky Way galaxy, which are the rough material for the formation of new stars. But the spiral arms are almost complete and already closer to the center. In contrast to the new ones, the arms contain more stars than dust.

While on their way out to the rim of the universe, the gravitation of the center mass shrinks the galaxies. They rotate faster, and the spiral arms finally close completely in. Now we call them elliptical galaxies. And the farther out, the smaller they are. (But the black hole in the center is now much bigger.) At the farthest distance we will find only small ellipticals and funny looking irregulars.

So far I have attempted to explain why the Big Bang theory is illogical. I never liked it, and many scientists do not like it either. As I already explained, the whole Big Bang business makes no sense, contradicts itself, and contains no proof of logical thinking or common sense. I am reminded of the following Jewish proverb: "When man thinks, God starts laughing." I hope He just smiles at my theories, that is fine with me. But I am very sure that He got a big laugh out of the Big Bang theory.

While we are at it, maybe I should answer a question you may have: How is it possible that I, a single person, can develop three theories of such importance, and why could I find out how and where Einstein made great mistakes? The answers are simple. I have the same brain as Einstein and all the others – human. I just use it in a logical way. And all the others made two great mistakes. First, they think on a much too small scale, and second, they do not include the cause of it all, the causal, cosmic intelligence, God. They do not know that the spiritual side of the universe is much more important than the world of matter.

Before I continue with my new theory, let us summarize what we have covered so far. The causes for the expansion of the universe are:

1. The whole universe rotates like a huge wheel, with all its members locked in location by balanced forces of gravity and magnetic fields. The only non-rotating movement is the journey to the outside of the universe. A strong vortex of energy penetrates the universe, forcing everything to follow the rotation.
2. The centrifugal force accelerates all galaxies to the rim. The farther out, the higher the rate of acceleration, caused by the higher rotation velocity which enters the multiplier of the equation for the centrifugal force squared.
3. While on the way out, each object gains on mass, which also is a multiplier in the formula, but not squared.
4. The growing radius is the only divider in the equation, slowing down the process, but is not squared. It cannot compete against the squared velocity in the multiplier.
5. During the journey to the rim, a black hole eats every galaxy from the inside until finally only an immense black hole is left.
6. The stars of a galaxy do not orbit around the center; they rotate, because a galaxy operates also like one big wheel. Galaxies do not expand; they shrink because here gravity is the stronger force, high density being the reason.

Here, I would like to add one more thing about the expansion of the universe. The trip to the outside of the universe will require a very long time; so long in fact that it exceeds the maximum lifespan of any star many times over. New stars are born within a galaxy all the time. Therefore, the stars we see in the most distant galaxies are about as old as the stars in all other galaxies, including ours. Only the black hole in the center is as old as the full lifespan of the galaxy.

It is a well-established fact that higher mathematics can take on a life of its own and leave the world of reality. A most primitive proof for this statement would be: 5 minus 8 = minus 3, which means three less than nothing. Right? Anyway, this reality-independent mathematics allows us to see the world as something consisting of ten *geometric* dimensions. Or, before the Big Bang, as an infinite small point with an infinite large mass, and being under infinite high pressure and temperature. Mathematics can indulge in a universe that has ten dimensions and then splits at the time of the Big Bang into a six-dimensional and a four-dimensional universe, with the four-dimensional universe being ours. The other one can be reached only by passing through a black hole. (So far there are four different theories about black holes, named after the mathematicians who spent years trying to understand them.) You think that all of this is nonsense? Of course it is! But leading scientists wrote serious books about these ideas (and got paid for it).

Why are they doing this? They believe that by the use of so-called super-mathematics it will be possible to unite quantum mechanics with relativity into a great-unified theory of everything (TOE). Einstein worked on it for years, and at the time of his death it was still on his desk, unsolved.

I think it is no wonder that God laughs about these ideas. The universe does not work by equations that cover a whole blackboard on the wall; it runs by logic and common sense! I do not believe that we can solve such problems by leaving the world of reality and escaping into a fictional world of super-mathematics. We need a picture we can understand and comprehend.

## Quasars and Antimatter

The most distant objects are the quasars (contracted from *quasi-stellar object*). Their location is odd. We find the first ones between 5 to 10 billion light years away. Within this range we also have the last larger galaxies. Then, from 10 to 15 billion light years away are most of the quasars, and the galaxies are elliptical or small and fragmental. Then, from 15 to 20 billion light years away, the region at the rim, we see only quasars at a very low density, very far apart from each other.

Quasars are extremely bright and emit energy, equal up to the output of a hundred regular galaxies. They are very compact, only about the size of a solar system. Most of the energy is infrared radiation, the rest are gamma ray photons. Quasars are also strong X-ray sources, and about 10 per cent of them seem to be a radio source. A massive outflow of gas has been observed from some of them, speeding away at thousands of kilometers per second. Many quasars are variable by a factor of two or more within a few hours. What are these power monsters?

Some believe that quasars are the violent nucleus of a galaxy-to-be. They (the Big Bang theorists) think so, because we see the quasars as they were 5 to 20 billion years ago. Others think the opposite, that quasars are galaxy clusters that collapsed into a violent nuclear furnace. But none of these theories can account for the gigantic energy output from such a small object.

My theory of the quasars is entirely different. It is based on a general theory of the cosmos that requires a very open mind and a good measure of imagination. Here it is:

The whole cosmos (spirit, nature, physics) has to obey the cosmic law of duality. Every force has a counterforce, every polarity can exist only with an opposite polarity, and so on. The list is endless. Because of this law, for every particle of matter, there must be an antimatter particle. Quantum physicists know that; it is a fact. But where is the antimatter? It cannot be in our universe, because matter and antimatter annihilate each other in a violent reaction when they meet. Such reactions release more energy than nuclear reactions, much more by magnitudes. In a bright flash the output is in the form of gamma ray photons. (Do you already hear "quasars"?)

In the beginning of this book I explained that the cosmos must be infinite; logic requires it. Because our universe is finite, this infinite cosmos must be filled with an infinite number of universes. And here I see the solution for the problem with the antimatter. Every second universe is an antimatter universe! We have matter universes alternating with antimatter universes, and enough space between. (I expect the inter-universal voids to be larger than a universe.)

Richard Feynman, the great American physicist, demonstrated in 1949 that antimatter particles have their time moving in the opposite direction as for matter. He had proof for this in the diagram, named after him the Feynman Diagram. Actually it is only logical that antimatter must have anti-time. If we extend these findings by Feynman to my picture of the cosmos, then all the antimatter universes have their time going backwards. That is from our viewpoint. But from the viewpoint of the causal cosmic intelligence, God, the two opposite time flows cancel each other out. Therefore the "official" cosmic time from His viewpoint is a permanent *now*! It is an extremely clever solution for having an infinite time condition.

These two opposed time directions are the actual reason why matter and antimatter cannot exist together and annihilate each other instantly. A particle or compound, made up from matter and antimatter atoms, is unthinkable. In fact it is impossible because of the opposite direction of time flow.

Before we leave the topic of the infinite cosmos, I must direct your attention to this thought as a logical consequence. Because of the infinity of the cosmos we must be ready to conceive the existence of universe groups and clusters, similar to the galaxy groups and clusters in our universe. This is the largest picture possible, and there is no limit.

As I have already mentioned, all galaxies end up as black holes when they are closing in on the cuter rim of the universe. As soon as they reach the speed of light, they cannot remain in our universe and must leave it by flying away in a tangential course. After a long journey, much longer than their lifetime was within the universe, they enter another universe, which is primarily a universe of the opposite matter. In the average of all universes about 50 per cent of all black holes escape in this way, the other half does not make it to the rim. What happens to them?

The black holes from the antimatter universes do the same, travel through the cosmos, and enter the outskirts of our universe. When such an incoming antimatter black hole meets one of our outgoing black holes, they find out that their linear time runs in opposite directions, which makes coexistence impossible. It is even impossible for the larger one to swallow the smaller one. In the end they annihilate each other with a tremendous outburst of energy: A quasar!

This annihilation occurs instantly by universal standards, but for us it can be anything from a few minutes to many years, depending on how they meet. In a head-on collision they meet at twice the speed of light, and we talk about a few minutes. Such a collision results in a short burst of hard gamma rays that last only seconds to minutes long. (Most measurements are around 15 seconds.) After a few minutes the whole story should be over, because the size of these black holes is not big, only a few light minutes across. But most collisions will not be head-on. The black holes may just touch each other, or meet tangential to the rim of the universe in the same direction. From there they meet their fate by first orbiting each other. Therefore many quasars will last from a few hours to many decades. We should observe many strange leftovers from the larger of such "meetings".

I mentioned the fact that the annihilation as a result of contacts between matter and antimatter result in radiation of energy that surpass nuclear reactions by magnitudes. How much? Nuclear fission (atom bomb) goes by Einstein's $E = mc^2$ ($c^2$ is 90 trillion). I believe the annihilation goes at least by $E = mc^3$. $c^3$ would be 27 quintillions (27 and 15 zeros). And while "m" at nuclear explosions is at the most the mass of a giant star (supernova), "m" in a quasar is the mass of at least two galaxies! That must provide for a fine "annihilation bomb", the quasar! This is my explanation of the astounding energy output of a quasar.

Of course, our scientists are thinking very much about the problem of why there is no antimatter in our universe. Let me introduce the latest idea: We can read (in *Smithsonian*, *Discover*, and other magazines) that during the first microseconds of the Big Bang, as the particles ran through their production chains, a tiny asymmetry ensured that for every billion antiparticles, there were a billion *and one* particles of ordinary matter. In a short time, all of the billion antiparticles annihilated with the billion particles, leaving one lone particle of ordinary matter. From that lone particle and its brothers the entire universe was constructed. You be the judge!

## Creation Through Annihilation

The output of quasars are in the form of gamma ray photons, and photons of other frequencies, which radiate into all directions. (Note: Photons have no anti-part and serve matter and antimatter alike.) Coming from all sides, they reach us here in the center of the universe, and naturally produce the highest density of this radiation. As a result, the highest number of collisions with other photons, particles, or bodies of matter happen here. The subatomic particles, required for the composition of matter, are so produced. These products of the photon collisions together with the output of supernovas and other particle sources are the raw matter for new worlds. (Stars of different sizes then breed the heavy elements.)

With the radiation of the quasars arriving from all sides, most collisions of energy quanta (also called particles) occur in or near the center of the universe; that is why most new galaxies form here. But, of course, much occurs also farther out. Matter is created in this way, and new worlds begin to form. First there are immense dust clouds. Then, at billions of points within the cloud, solar nebulae will form. These solar nebulae condense farther until a star ignites at the center, and planets begin to form. Because of the whirlpool effect within the rotating force fields of the universe, this whole gigantic cloud of nebulae and stars begins to rotate. Following the laws of physics, spiral arms will form and this new galaxy will shrink and rotate faster. (Here we can see that orbiting is not possible because there is nothing much in the center to provide the required gravitation. A galaxy begins like one huge wheel and remains that way. Ditto for the universe.)

Then all these galaxies begin their journey to the outside of the universe in order to fulfill the eternal cycle of creation and annihilation; the principle of the Tao! The picture of the universe, as we had so far, shows clearly that the cosmos is not something that occurred by chance. When we see how everything fits together, how cleverly it is designed, we must realize that only a superior intelligence could create such a marvelous thing.

How old is our world? From my perspective there is no age for the universe; it is eternal. But the components of the universe (galaxies, stars, etc.) do have a lifespan that is limited; they have an age. My estimates are these: Our galaxy seems to be approximately 15 billion years old. The galaxies of the center, the Virgo supercluster, could be around 6 to 10 billion years old or very new. On the other hand, the farther out the galaxies go, the older they are; the galaxies, not the stars. Stars are born and die all the time. This makes it practically impossible to measure the age of a galaxy in any way. We already know that the black hole in the center is the only part of a galaxy that has the same age as the time of existence of that galaxy. And there is no way to measure the age of a black hole; we are lucky if we can locate one indirectly.

All of the above does not apply to the very few quasars which exist inside the 5 billion light years radius, like TON256 or Q500241. They are brand new. I think they are caused by antimatter black holes that came in from the flat side of the universe and collided with one of our galaxies. Such collisions should result in very strange looking quasars, surrounded by stars. These kinds of quasars are scarce, because the way through the universe from the flat side is very short, so most black holes just zip through without hitting anything.

Creation and annihilation go hand in hand and keep the universe in a state of permanent existence. There is absolutely no need for that funny Big Bang. This principle, as just explained, keeps all universes going forever, and equally serves matter and antimatter universes, with their anti-galaxies, anti-stars, anti-planets, and anti-people. Our matter is antimatter for them!

Before I move on to the next topic, I would like to answer a question that some readers may have: By analyzing the red shift, we detect the escape speed of objects in the universe. So why cannot we see the side movement of the orbital speed?

Actually we cannot talk about orbits around the center of the universe. The whole universe rotates like one huge wheel, just like the galaxies do. When we look out into one direction, all objects along our line of view are on the same "spoke", so to speak, and therefore they seem to stand still. We rotate also on the same "spoke", but at a lower velocity, because we are closer to the center. The only way to detect the side movement of the rotation would be to see another universe in the background for reference. Maybe someday we can do that by installing a huge telescope on the moon, but I doubt it will work.

Every month we can read in our scientific magazines about new ideas and theories that deal with problems of the universe, most of which are written by leading scientists. Each one on its own does not look too bad; sometimes they are even convincing. But if I list them one after the other, we can see that science is not only fun, it can also be funny. (I cannot explain each one in detail, and I do not want to, because I believe they are all wrong anyway.) So here is the list that hopefully will enable you to judge better what you read or hear:

1. White holes, the opposite of black holes, where matter appears spontaneously out of nothing.
2. Wormholes that connect a universe with a baby universe, or two regions of space-time, a fine mesh of space-time-loops that are the basic fabric of space and matter. Others do not like that and postulate space-time-strings as the stuff of the universe.
3. Doughnut-shaped black holes, where you, with some luck, pass through the center and reappear in another universe at another time.

Each one of these (and many others) is taken very seriously, and even fancy books are written about these ideas. I produced the above list only to demonstrate why I can say that my theories are based on common sense.

## Spirit and Matter

A very good and also very old question is: How does the mind of the causal cosmic intelligence, or God, keep control over every single particle and every galaxy in the universe? We can find a clue to the answer in unexplainable results of experiments in particle physics, known as quantum mechanics. Experiments in particle accelerators showed that particles can indirectly interact in a superluminal fashion (faster than light, instantaneous). The experiments showed that subatomic particles somehow communicate with each other. They seem to know in advance what the other will do. There is no explanation for this so far. I believe that with these experiments the scientists hit the borderline between the physical world and the universal mind. And I am sure that here is the end of our physical possibilities. The thought energy of God is here at work; we even have a name for it: Spirit!

It is the spirit that does the trick. There is no way for us to sense or measure spiritual energy; it is not any kind of energy we know or can control by physical means. It is the mind, or thought behind all things. In fact, the cosmos is mind-stuff, even down to the smallest subatomic details! The so-called quarks and the other subatomic particles are monsters in comparison. We will never learn the true nature of spirit, as long as we are with our consciousness in the physical realm of this low-level planet. But these experiments can give us some idea of how God controls everything, from subatomic particles to galaxies. Maybe the following could be of help in understanding the situation: Everything and every thought in the universe is like a cell in the body of God. Just as every cell in your body knows what it has to do, and as it follows the orders of the body's mind, so it is similar with everything in the body of God.

I think that this illustration shows better than anything else that we are really created in His image. All matter, all energies, all forces, the whole spiritual realm, and every thought originated with God and *is* God!

In Chapter V I explained what the ethereal beings of the spiritual realm are doing, and what some super-civilizations are able to do. They use the most powerful force of the cosmos, the spiritual thought energy, to manipulate the evolution of the physical universe. Therefore, when we observe an event in the universe, and our laws of physics cannot explain it, then it may be manipulated by the power of spirit.

When anyone thinks that I am a little bit far out with all of this, he should study Bell's Theorem and the resulting idea of super-determinism. It is definitely one of the best ways to get completely confused about reality.

The next topic will be a very interesting one. But before I enter into it, just one more thought about our solar system. When the comet Shoemaker-Levy was caught by Jupiter, the result was a very long stretched elliptical orbit, with Jupiter as a focus point at the far end. Exactly the same happens to most comets who orbit the sun. They have the same kind of very long, stretched ellipse, with the sun as the focus point at the very end. The other end reaches far out into space, far beyond Neptune and Pluto. The fact that most comets are dirty ice balls gives me this idea: Space is filled with lots of hydrogen atoms, and also with plenty of oxygen atoms. They are all subject to the gravity of our sun. As soon as they are close enough, they react and become water (or rather ice). The ice particles combine into larger bodies and begin a journey towards the sun when one of the outer planets disturbs their orbit. Scientists are perhaps right in thinking that comets were the source for the water on our planet. The earth was able to keep it. The others got the same, but on Venus it evaporated and oxidized away; on Mars it oxidized into the rocks; and on our moon it evaporates always within 15 days (it rotates once in 30 days). The comets of today are only leftovers. I have a feeling that at the beginning of the solar system it rained comets.

## Space Travel and Extraterrestrials

One problem is still not solved: Can we travel through the universe, or at least within the galaxy from one solar system to another? And how can we do it? Even after one hundred years of technological progress, our so-called spaceships are as far away from this capability as the plane the Wright brothers built was from a Boeing 747. In short, our latest rocket technology is still something very primitive. But even if we had the highest possible physical technology, if we could travel at the speed of light, we can forget about travel into deep space to other stars. The vast distances to other stars and our short lifespan are forbidding!

A very short round trip to the nearest star, Alpha Centauri, would require 9 years, going at the speed of light. For other destinations the distance would mean 30 years, 500 years, 2,000 years, or just 2.25 million years to the nearest galaxy, the Andromeda system, that is, if we travel at the speed of light at 300,000 kilometers per second. (A round trip to Andromeda would be 5.5 million years!) I am very sure that we would never survive even the short trip of 9 years. At the speed of light we would be hit by space debris with absolute certainty. Even a pea-sized pellet would result in total disaster at this speed, destroying the ship. So we had better forget about physical space travel out of our solar system to other stars. Of course, within our solar system space travel is no problem, as long as it is at decent speeds. What the upper speed limit for safe travel will be, we have yet to find out. But one thing is sure: We have to develop something much better than our primitive fuel-powered rockets!

But deep space travel is going on right now all over the whole universe, and traffic is heavy! And the travel time for all distances is the same: between zero time and a few seconds, even across the whole universe. I am sure this makes you laugh. I know that you cannot believe such nonsense. But this book contains no science fiction. Please keep an open mind for what is coming; it is very interesting.

If you belong to the group of lucky people who have had an out-of-body experience, then it will be very easy for you to understand the following, because you have already experienced change of location in no time at all, and you know it is real. But even without such experience it should be easy to conceive.

When we are in our natural ethereal state, just a pure soul and not incarnated in any kind of body, we can travel freely and instantaneously to every destination by just thinking of it.

In this state we are pure conscious thought energy as an individualized part of the thought energy of God. This kind of travel is accomplished in three steps, which are in fact the same steps we do for every kind of trip here on earth. First, we decide which destination we want to go to and why. Second, we desire to go there. Third, our free will gives the command to go. In this instant, we disappear from our old location and pop up at our destination. Instantly! Each of us did this before we first incarnated here on earth; it is normal. And we still do it very often between incarnations, when we have to go to other planets or solar systems where we will have special activities and learn lessons that are required by the nature of our karma – lessons that cannot be learned in any other way.

This kind of travel by individuals is going on all the time all over the cosmos. It is the easiest way to go. But there is another kind of space travel that is more impressive to us, because we live and think in a three dimensional world. Many human (and non-human) inhabited planets in the universe have super-civilizations with super-technology. They live in complete contact and harmony with the spirit of God, and their mind is at all times permanently spiritually oriented. They are always fully aware of who they are. These civilizations have complete control over their planet and solar system. Only people in perfect attunement with the God-spirit are allowed to live there. These are the people who get assignments from higher entities of the universal hierarchy to take responsibility for the development, control and maintenance of new planets. They seed the required life forms, correct planetary conditions, etc. These people are able to do the following with a technology that operates on the borderline between the physical and spiritual. In fact, they combine the two. They built small and huge spaceships, which manipulate the omnipresent gravity (next chapter) for power within an atmosphere or for short trips of a few light minutes or hours within solar systems. To make interstellar trips (to other stars or galaxies), they change the whole ship into an astral-ethereal state, and then, by just thinking the three steps, they go instantaneously to their destination, just as individual souls do. At arrival, they re-manifest the ship, and everything in it, back into the physical state and do the planned task.

That is space travel! It is the only way to fly out there! Nothing else will do, not even an *Enterprise* at warp speed. And of course, the "beam me up" business as in *Star trek* is not possible. Via spirit is the only way to do that.

Maybe you still cannot believe all this. I think that the

following could be of help: There are many cases of teleportation here on earth that are well investigated and documented, many by the Catholic Church (see the book *Miracles, a Scientific Exploration of Wondrous Phenomena*, by D. Scott Rogo, for example). But I was lucky to have had contact with one case of teleportation in 1974. We had hired a contractor to do some work on our house. During the week that he needed to do it, we learned that he was a very unhappy man, because nobody believed him and everybody laughed at him. He trusted us and told his story. I cannot give his name; he may still be alive.

He was on vacation, with his family, at a seaside hotel in Pebble Beach at the California coast. They were sitting side by side in front of their room to enjoy the sunset. He thought of home and in an instant he found himself at home in La Habra, near Los Angeles. He called his wife, who was in panic, at the hotel and promised to return. Then his son drove him back to Pebble Beach. Because we were the first people who believed him, and also could give an explanation, we were his dinner guests two weeks after he finished the job for us. His wife told us that this terrible second, when he disappeared in an instant, will haunt her to the end of her life. Two years had past since then, but she was still very touchy about this event.

Our explanation was simple. His subconscious mind did this without him being aware of it. He was lucky to end up at home, and not somewhere else.

Returning to our basic topic, I would like to cover another very popular question. Is time travel possible? Yes, definitely! But not in any way our scientists are speculating about it (using black holes, cosmic wormholes, etc.). These beings that are able to travel through the universe by using the spiritual road can do it. Because there is no linear time in the spiritual realm, they can not only set a destination, but also at what time they want to be there; past or future, is immaterial.

Now we have arrived at the topic of UFOs. From the thousands of UFO sightings, at least 90 per cent, if not more, can be explained as something normal or as hoaxes. But that leaves us with hundreds of cases that we have no explanation for. They are real! They are the ones we have just talked about. And now we also understand why in so many sightings the ships appear suddenly in the sky, and even more often, disappear in an instant. I share the opinion of many authors that there is no more doubt that UFOs exist. There is more than enough proof worldwide. So why do our scientists shy away from this problem? They should do exact scientific research in order to find out what they are. I guess they are scared of what they may find.

But why do we have so many visitors right now? That is the question. Who would be interested in visiting a tiny planet of our sun, which is an unimportant star at the outskirts of a galaxy with over 100 billion stars? I think these visitors are the people who are responsible for our planet, as I explained a while ago. Because the earth is under permanent observance by them, they recognize that we are making a mess out of this planet. They do not like to see their hard work ruined by our ignorance. That is why they are intensifying their observations right now. They try to correct whatever they can, but by cosmic laws they are not allowed to interfere with our free will. That is why it is important that our scientists not only find out who they are, but should also get in contact with them. Then we can ask them for assistance and help.

So they will continue to watch us, and someday they may have to interfere in some way. We cannot even guess what they may do. The mental difference between them and us is at least as big as the difference between us and apes, probably bigger. (Mainly from the spiritual viewpoint.) For just this reason we cannot expect that we will be contacted by them in any official way; we will never be able to catch them. They are too superior. And being so spiritually advanced, I believe they are members of the spiritual hierarchy of this section of the galaxy. I am very sure that they will not contact us. They will wait until we are ready and worthy to join them in their federation when *we* contact them.

There are dozens of books and videos that have investigated into past landings and contacts with these people. It is said that they supplied their knowledge and technology to Atlantis. And the Sumerians left it even in writing that "people from the sky" gave them all the knowledge, seeds for grains, and even intermarried with them. But I am very sure that they will not do so today because of our propensity to shoot at everything we do not know or comprehend. These extraterrestrials (ETs) have complete knowledge of what and how we are, and why we are incarnated here on earth. They do not interfere because they want us to learn the hard way on our own. All they can do at the moment is influence our minds in order to get us back to our senses – and they really do it.

We have had radio for about hundred years now, and lately even radio telescopes. Have we heard any artificial radio signals from outer space? According to writings in *Discover* magazine and other scientific publications in the 1980s the answer is yes and no.

Yes, the scientific publications talked at that time about six signals from outer space that were definitely artificial. They came from six different locations in the galaxy. None of these signals came from sources on the earth, satellites, or space probes.

No, said the scientists and did not release any more data after these six. Why? Because each lasted only a second or so and were not repeated, none of them can be counted by science as a proven fact. This is the way, our scientists think, and it is absolutely right. If a test cannot be repeated or duplicated by others, it is not science. Only experiments that we can duplicate, count. But there are, in my opinion, exceptions.

The following is a clear-cut example of how scientists can carry this principle to excess: In 1997, there was a reader's query in *Astronomy* magazine. The reader wanted to know why scientists shy away from the topic of UFOs, even though there are thousands of sightings. Scientists should investigate. The answer of an astronomer was as narrow-minded as humanely possible. In essence, what he said was that sightings have nothing to do with science because they cannot be duplicated. Period!

So, why should these signals be an exception? I think it is a special situation, and I hope that scientists will someday agree.

The strength of a signal from an omnidirectional antenna, like our radio and TV stations, is inversely proportional to the square of the distance from the antenna (same as Newton's law for gravitation). Or in simpler words, as the distance gets greater, the signal gets twice as weak with each step as before. There is no way that anyone in our galaxy will ever receive our broadcasting signals. They diminish like the waves caused by an acorn that we throw into one of the great lakes. Directional antennas, as we have on radar or space probes, are a different story altogether. These antennas concentrate all the energy in a very narrow beam, and so reach farther by magnitudes.

Now let us imagine the following: If a directional antenna on a planet far away in the galaxy sends a radio beam out into space, maybe to communicate with a spaceship or probe, most of the energy will bypass its target. And this beam *may* accidentally hit our earth for one second or a little more. But it will not happen again because that other planet rotates and orbits, as does its whole solar system around the galaxy. Our earth does the same, rotates and orbits. Our scientists should accept each one of these very short signals for real, using this reasoning. And if we consider the extremely low probabilities of such accidental hits, there must be thousands or even millions of directional antennas in operation out there. In my opinion, the chance of hitting us accidentally twice are less than the odds of winning the jackpot in a lottery, squared.

Some scientists speculate that maybe people on another planet have a huge directional antenna in operation, permanently aimed at the earth. I cannot believe that anyone out there spends money, effort, time, and energy for a powerful directional antenna, and is willing to wait hundreds of years for an answer. For just this very reason most scientists do not want to spend money and energy on such a hopeless project.

I think it will be best, if we let common sense and logic succeed, and accept the signals we receive for what they are: A very rare, accidental glimpse at another civilization out there! Scientifically proper duplication is practically impossible, so let us face it as another exception of the rule.

There is one more thing we have to consider. These signals can only come from civilizations similar to ours, using the same means of communication as we do – modulated electromagnetic radiation. But how long does such a period last? Let us look at ourselves. After thousands of years of civilization without electricity, we have now worked with radio waves for about one hundred years. I believe we will do so for another 500 to 1000 years. Then we will communicate via spiritual channels (telepathy, etc.), just as the super-civilizations do. To operate within the universe, the radio waves are too slow because they are limited by the speed of light. The super-civilizations must use superluminal means (faster than light), based on a physics of spiritual energy which works instantaneously.

I mention these factors only to point out that we cannot intercept the signals of most of the higher civilizations because our infant "high" technology of only one hundred years is behind these civilizations by thousands of years. They cannot even introduce us to their technology; it would be beyond our comprehension. They know that and therefore do not do it. (We would use it for our wars anyway; they know that too.)

One more question is in the mind of all of us. How do these people of the super-civilizations live on their planets? The answer may be surprising. As a result of years of study I have arrived at the following picture:

They live a comfortable life in houses that are just large enough and disturb the nature around it very little. Their planet is covered as little as possible with concrete, roads, and big cities. Traffic is almost entirely by air in machines that use gravity control for power. Most personal travel is by teleportation. In order to leave the nature of their planet untouched, they have their mining and industries on the moons or barren planets of their solar system. Manufacturing that must be done on the planet is underground whenever possible. These people live in harmony with the nature of the planet, not hypermodern as in our science fiction movies.

Albert Einstein wrote in his book *The Evolution of Physics*:

> Fundamental ideas play the most essential role in forming a physical theory. Books on physics are full of complicated formulae. But thought and ideas, not formulae, are the beginning of every physical theory.

This chapter contains my theory about the universe, which is my "free invention" as Einstein liked to say. The only way to prove it to be wrong is, I believe, by using opinions and dogmas. It solves just too many current problems in cosmology to be completely incorrect.

The next chapter will also be mainly science. I would like to remind you of the purpose of this book. This book provides spiritual knowledge for the science-oriented reader, and science for the religious-oriented reader, in order to get a complete picture of the universe and ourselves.

# Chapter X
# A THEORY OF EVERYTHING OR "QUANTUM-RELATIVITY"

> When the solution to the big picture comes, it will be so simple, so pure, and so obvious that we will marvel at ourselves for not having seen it all along.
>
> John Archibald Wheeler

Most scientists are dreaming of the famous "TOE", the theory of everything. They believe that this has to be done by uniting the theory of relativity with the theory of quantum mechanics, and of course, they see it mainly as a mathematical problem. Albert Einstein was one of them, and he spent the last years of his life working on a solution. He did not succeed, and all his work was still on his desk when he died, unsolved. I believe he could not succeed because he made the same mistake as all the others. He tried to crack this hard nut with sophisticated mathematics, and he did not include the cause of it all – the spirit! The world is not equal to one huge mathematical equation. A theory of everything must show a picture of everything there is! Mathematics is only a tool to handle the situation. Relativity and quantum theory are only a part of the world. And if we look very closely at these theories, we see that only quantum mechanics and Einstein's $E = mc^2$ (which is in fact quantum business), are the only ones that provide a picture of things. Relativity is only a way to look at things, but it does not say anything about what we are looking at.

The entire content of this book provides the picture of a theory of everything. Einstein and all the others could never solve the problem because they do not accept the existence of a causal intelligence, or God. Without the causal energy, also called spirit, the picture will never be complete. A complete picture of everything must include the causal intelligence, the spiritual world of the cosmos and its laws, all energies and forces, and the physical world of matter. That will be everything, not fancy equations! And this book is trying to provide just that. This chapter will introduce the last theories and facts to complete the overall view, from which you can then draw your own opinion.

After this introduction to the TOE, a few words about the subtitle of this chapter. Quantum and relativity are two big words and many people are scared of them. Believe me, there is no reason to be scared. All the books that deal with these theories are filled with a lot of special details, even hairsplitting that are useful only for the scientists involved, but of no importance for us to understand the theories. So, if you have so far believed that quantum theory and relativity is something too high and complicated, understandable only for a genius, then please relax. Albert Einstein recognized this problem of so many people too, and he wrote a book, *Relativity*, for every reader who had a high school education. It is a very good book and makes for easy reading.

Now I will give a short introduction for these two theories and the problems we still have with them. I will keep it as short as possible.

The German physicist Max Planck devised the quantum theory in 1900 in order to account for certain phenomena that could not be explained by classical physics, namely the radiation of a so-called black body. He postulated that electromagnetic radiation consists of *quanta*, the word he coined for these kinds of subatomic particles. When Ernest Rutherford discovered the structure of the atom, the new quantum theory proved to be very handy for the understanding of the charge-paradoxes of the atom, Niels Bohr developed these principles further, but it became clear that the quantum "theory" was a fundamentally weak answer to all questions about the atom.

Beginning in 1925, quantum "mechanics" was developed to take the place of the quantum theory. Louis de Broglie, Erwin Schrödinger, and Werner Heisenberg are the physicists who did most of the basic work. Quantum mechanics states that all subatomic particles have a wavelike nature, actually quantities of energy. A branch is Schrödinger's wave mechanics, and so is Heisenberg's uncertainty principle, which states that nothing is certain and that we are dealing only with probabilities. For example, it is not possible to know with certainty the position *and* momentum of a particle. Only one of the two can be known with accuracy. When Einstein first heard about this, he did not like it at all and made the now classic remark, "God does not play dice!" But later on he accepted quantum mechanics and even contributed to it. (Without quantum mechanics there would be no sunrays and no atom bomb.)

Better known and more often talked about is Albert Einstein's theory of relativity (there are two). In 1905 (at the age of 26) he proposed his theory of special relativity. At that time the physicists laughed about it. Now it is a kind of religion.

Galileo Galilei (mostly called just Galileo, which was his first name) developed a theory of relativity, known as the laws of relative motion, hundreds of years earlier. For example, if you walk at 3 miles per hour in the forward direction on a train that moves at 60 miles per hour, then you can see it in two different ways. You walk at 3 miles per hour in reference to the train, but you move at 63 miles per hour in reference to the track. Einstein "improved" this law by adding electromagnetic radiation (light, radio, X-rays, etc.), which was unknown to Galileo. (Later we will see that this was not really an improvement.) Einstein's special relativity deals with the conditions that different observers will find when moving at a *constant* speed in respect to each other. (Accelerated speed is the topic of his other theory.) A whole, thick book could be filled with all the possibilities of this, the example of the train as given above being one of them.

But the main reason for developing this theory was something else. Before he became a physicist, it was already proven by many experiments that the speed of light is always the same, under all possible circumstances and frames of reference, and is totally independent of the relative motion of the observers. In other words, light does not behave according to Galileo's law of relative motion. It is always the same, 299,792.454 kilometers per second, ±zero in a vacuum, generally stated as 300,000 kilometers per second. This was a terrible fact, absolutely contrary to the established laws of physics, because it violates the classical concepts of relative motion.

Einstein saw his life's mission in solving this problem and went to work on a theory that will explain everything. He thought that this light problem must become part of Galileo's relativity. He built his theory and in the end made the bold statement that mass and energy are the same and then introduced to the world of physics the famous $E = mc^2$, which means that energy equals the mass (m), multiplied by the speed of light (c), squared. This formula proved to be right in many ways.

He also proposed, based on the equations, three conditions. First, when an object reaches the speed of light, its mass becomes infinitely large, and no energy in the world can accelerate it any faster. This is the reason why the speed of light is the absolute upper limit for everything. Second, at the speed of light an object will have zero length, which means it ceases to exist. Third, at the speed of light time does not exist anymore for the object, time stands still.

All of this is valid only for an outside observer, but not for an observer inside the object. (In Chapter IX I have already explained why I cannot accept these paradoxes. I will go deeper into the why later on in this chapter.)

In 1915, Einstein presented the general theory of relativity, which is of importance chiefly to astrophysicists and cosmologists. It brings together acceleration and gravitation, stating that an accelerated object is equal to an object under the influence of gravity. Both seem to be the same, or at least have the same effect. They can add to each other, subtract from each other, or cancel each other out, if they are equal (like a Satellite in orbit). The general theory also states that because energy is equal to mass, gravity can bend a beam of light (proven many times in tests). It also explains gravity as the product of curved space-time caused by a mass of matter (purely mathematical).

This should be sufficient for the reader who is not familiar with these theories. More is not necessary to understand the contents of this chapter.

As I already said, Einstein worked through the final years of his life on a solution for the so-called grand unification, the problem of uniting quantum mechanics with relativity. One thing he found out for sure was that gravity is the hardest nut to crack! He cracked it somehow, but the result was an illogical mess nobody can comprehend (similar to a dog chasing its own tail). After Einstein, hundreds of scientists tried to find the solution for the grand unification; but so far all of them have failed. Some believe that they solved it with exotic super-mathematics, but without any proof.

I believe that using only mathematics is the wrong way to find a solution for problems of such cosmic proportions. The only right way I can see is to follow the method of the classical thinkers. First, there must be the idea, the theory, or intuition. *Then* we can apply physical laws and mathematics in order to get a grip on the new theory. (Einstein believed in this way of doing science too, and he did always start with an idea. But the idea must be right, and in the case of his relativity theories he had the wrong idea at the very beginning around 1900, as we will see later on in this chapter.)

But most importantly we should not repeat the cardinal mistake Einstein (and all others) made: Not to include the source of it all; causal cosmic intelligence, or God, and the energies and forces of the spiritual realm. Without including these basic factors we cannot arrive at a final solution for the problems of the universe.

Of course, I have tried to find a solution too, and I believe I have got the right ideas. Why not? I am not a scientist, but I can think like one because my brain is exactly the same model as theirs – human. As a young man it was my goal to become a scientist but circumstances, primarily the Second World War, locked me into engineering for my whole life. But I am free to be an amateur scientist. There are many amateur astronomers, using their backyard telescopes every clear night. They have made great contributions to astronomy, and the astronomic community is thankful for their work and honors their findings.

My problem (in being accepted) is that I am not an observing or experimenting amateur, but a theoretical one. Why do ideas and theories from amateurs have a rough time being heard? Is it the ego of the experts? Maybe.

Anyway, all I can offer is an idea and a theory, based on intuition, logic, and common sense. The development of the detailed physics and mathematics I must leave to some open-minded scientist, who has the courage to give these ideas a try. And I would like to repeat once more *why* everybody has failed so far on the job of the grand unification. They reject everything that smells even a little bit of spirit, metaphysics, soul, religion or God. I see this attitude in almost every scientific book or article, sometimes spelled out plainly and directly in one of the introductory paragraphs: "It is not scientific!"

## The Grand Picture

And now I present my picture of the world as I see it. Please keep in mind that all the previous chapters, one through nine, are an integral part of the whole picture. I have gathered the many items over a period of three decades of research, then added my own intuitions and ideas, and organized everything into one complete picture. It is not the grand unification theory. Unification of quantum mechanics and relativity alone will not do it (and is not possible as we will see). There has to be much more! It is like this: You cannot get a complete picture of the earth from a small island in the ocean. You have to get far out into space, and get a look from there. And even then you see only one side, so you will have to wait until the earth makes a complete turn to see the whole earth as it is. Equations and theorems do not suffice, a grand viewpoint is essential!

Did you know that a bumblebee should not be able to fly? That is right! If we apply all the formulas of the airplane industry, it can be proven that its body is too heavy for its small wings and its means of operation. It is definitely impossible! But the bumblebee does not know that, and because of this ignorance it flies anyway. Why do I mention this? Because I am an amateur scientist and I hope that my theory, just like the bumblebee, will fly; and that the scientists will see it.

Let us begin. First I shall present a very short overview, then the details. In the beginning of the previous chapter I explained that I use known words with a new meaning. The cosmos is the whole creation with an infinite number of finite universes, and universe means any of the matter or antimatter universes, including ours. This alone would already serve for the term "grand" in my picture, but it is only the physical world of matter. The really grand picture is very much more!

For a clear picture, a theory must be organized in a certain way. Some prefer to group these kinds of categories into different dimensions, others like the term realm better; I like to group by name. Actually, it does not matter; what really count is the message, and not the wording. (The same applies for the Bible!) I have six groups:

1. *The universe of matter*: Time, space, atoms, compounds, planets, stars and galaxies.
2. *Life of the universe*: The why and how of all organic life

forms throughout the universe.
3. *Energies and forces*: Electromagnetic radiation, electricity, kinetic energy, inertia, potential energy, heat energy, gravity, acceleration, and all the vector quantities.
4. *The Astral world*: Parallel and similar to the first three physical groups, but of an entirely different structure. It is also physical in a way, but not detectable from our physical world. Its existence is required by the law of duality.
5. *The ethereal realm*: Also called the etheric, it is the realm of the souls and free spirits. It holds our long-term (eternal) memory, emotions, free will, desire, etc. It is also the location of the *Akashic Records*, the cosmic/spiritual memory system.
6. *The causal, cosmic intelligence, or God*: Reigns over everything and *is* everything. His thought energy is the most powerful energy of the cosmos. All other energies are only one or another form of its expression. He causes everything to be. He cannot be comprehended or imagined from our third-dimensional physical level of existence.

And now let us go into the details of each one of these six groups, which represent everything there is.

THE UNIVERSE OF MATTER

What was the first step in the creation of the physical cosmos? *Time* and *space*! These two are the basic, mandatory requirements for the creation of a cosmos. Physical universes cannot exist without time and space. From the previous chapters we know why it is required. By using simple logic we know that both must be infinite, and cannot be comprehended from our level. Sources like Edgar Cayce, Paramahansa Yogananda, etc. say the same. Without space and time there would be nothing!

Another fundamental requirement is the building stones of matter. We call them subatomic particles. Actually, they are not particles at all; they are manifested energy in vibrational or wave-like form. And they are not the quarks, bosons, leptons, baryons or mesons, just to name the major groups that are considered to be the building blocks of matter by our physicists today.

These quarks and friends are giants, huge monsters, when we compare them with the real basic units of matter. The basic "particles" could be understood as the opposite end of infinity; they are infinitely small. They are, in fact, the smallest possible units of the thought energy of the cosmic mind, God. (Paramahansa Yogananda coined for these smallest units of the causal thought energy the word "Thoughtrons", a good name I think.) We cannot conceive, sense or measure them, but at least we know that they are raw building material *and* the controlling force for all physical and astral matter.

Millions of these basic units are arranged into certain shapes and represent an atom. Each atom is a real body in space, and subject to gravitation in the same way as rocks or planets, as we will see later on.

So far everything was very abstract and not easy to grasp. More familiar is what we call the manifestation of matter. Protons and neutrons are organized to make up the nucleus of atoms. Vibrating, oscillating or rotating (we are not sure) shells of charged energy units – the electrons – complete the atoms. The causal energy of the cosmic mind is very strongly involved in balancing all forces that work on the atom. How all of this occurs, we will see later, when I explain my theory of gravity.

Classified by the number of protons in the nucleus of an atom, there are 92 natural elements in a stable state. The smallest is the hydrogen atom with just one proton (#1), and the heaviest is the uranium atom with 92 protons (#92). Under normal conditions they are stable forever. The heavier so-called transuranium elements (higher than #92) are all artificial and not stable. They decay by radioactive radiation, the higher the number, the faster they disappear. Nature has no use for them.

When these 92 varieties of atoms combine into an uncounted number of molecules, and then molecules combine into compounds, an infinite number of substances are possible. And each one of these substances changes its properties under the influence of different temperatures, pressures, and other factors. This whole creation of physical matter is so fantastic that we are unable to fathom it.

As of today, our physicists have already gotten a large zoo of subatomic particles together. They have almost run out of names for them. Some of the results are useless for any application, but we have also learned very much from these experiments with particle accelerators (cyclotron, linear accelerator, synchrotron, etc.). So far we have arrived at this basic knowledge: Five subatomic particles are natural and stable; the photon, the electron neutrino, the electron, the muon neutrino, and the proton. The photon is the only one that has no antiparticle; it serves matter and antimatter. The other four all have an antiparticle. A sixth particle, the graviton, is only hypothetical and I am sure it does not exist. All the others are the result of experiments in accelerators and have a very short lifetime; one, $Sigma^o$, has only 59 sextillions of a second to live. One sextillion is a 1 followed by 21 zeros. I believe that all of these artificial particles are very interesting, and we may learn something from them, but they are useless for the construction of physical matter. I doubt that anyone of them is a building block for matter. Instead, I think that we are studying fragments of smashed particles, like studying the pieces of a shattered light bulb, and then classifying them.

So far, physics talks about four forces which control atoms and matter in general. Within the atom we have the strong, the electromagnetic, and the weak force. The fourth, gravity, is only used for the macro-cosmos (universe), and neglected for the subatomic world of quantum mechanics. (I think this is wrong according to my theory of gravity.) Some physicists already have the suspicion that the electromagnetic and the weak force are only theoretical variations of one and the same force.

# A THEORY OF EVERYTHING OR "QUANTUM-RELATIVITY"

*About the Universe:*
Of course, the most visible manifestation of matter are the planets, moons, stars, and galaxies of the universe. They are a direct result of the endless possibilities of the 92 elements, and so are all life forms in the universe.

There is one thing I would like to mention within this section about physical matter. It is an idea that will just not leave my mind. For eternity, an infinite number of stars have been very busy with their nuclear fusion; producing helium out of hydrogen. The solar wind of most stars contains not only protons and electrons, but also helium. Protons and electrons will always find a way to react and combine with something else. But helium is a noble gas, and has no affinity to any of the other elements. Based on what we know from our solar system, the cosmic abundance of helium in interstellar space (between stars) is estimated at about 28 per cent. (The rest is mainly hydrogen.) My idea is this: The corona temperature of our sun is 2 million degrees C (the surface has only 8,000°C), and most stars are much hotter. The helium radiates away from the stars and has to pass through this hot corona. I believe that the helium atoms will be ionized during this passage, which means they are robbed of one electron. Maybe someday we will have a way to find out if the interstellar helium is ionized.

As we can see, the universe is very interesting. It is the largest manifestation of physical matter we know. So, now we shall spend some time considering the latest findings and problems in cosmology.

With each new series of pictures from the Hubble Deep Field (HDF) observations, we do get some answers, but new questions pop up in even greater numbers. The density out there appears to be high, but most astronomers already see that this is an illusion as an effect of the tremendous depth of billions of light years. What we see are all the objects behind each other, and because we cannot measure these distances exactly, they appear to be close together. Most cosmologists still believe in the Big Bang (they have nothing better), so they consider everything out there to be very "young" and are confronted with a few problems of their own making.

When there are so few and such small galaxies, how did they develop into the many giants of today? Why are there only ellipticals and irregulars, and no spirals? How is it possible that these most distant galaxies are so mature, if they are new and under "development"?

Again and again astrophysicists measure the age of the universe as about 8 to 10 billion years, while the age of many stars is known to be 14 to 16 billion years (based on the Big Bang theory). Because children cannot be older than their parents, something is wrong, very wrong!

Of course, there are already some people who see the problems in a proper perspective. For example: In an article, written by M. D. Lemonick in *Time* magazine (November 7, 1994), we can read:

> The scientists' method of calculating the universe's age is based on the assumption that it has been expanding ever since the Big Bang. If the age estimate is wildly wrong, then there could be a flaw – possibly a fatal flaw – in the Big Bang theory... But if the age appears to be more like 8 billion, then the Big Bang may be shot.

As we can see, there is some hope. I believe that I can offer the required new theories. In the previous chapter I defined my new theory of the nature of the universe. For gravity and speed of light (in this chapter) I just turned everything 180° around and upside down, and it works. That was my opinion, but Mark Twain had one too:

> The researches of many commentators have already thrown much darkness on this subject, and it is probable that, if they continue, we shall soon know nothing at all about it.

LIFE OF THE UNIVERSE

Now we will discuss the so-called organic life forms. But what does "organic" mean? For us it is all life that is based on carbon and oxygen, but this pertains only to the earth. I could imagine other organic systems, based on some other of the 92 elements; we have some already on our planet. Many hot springs contain bacteria that are sulfur based. So, it is possible.

But let us talk about what we know, life on earth. All these life forms are designed and created by beings of the etheric realm, including us (that is, our souls). The religious reader may now protest: "But the Bible says that God created man!" That is right, actually He did, but we have to consider two facts. The Bible also says that God created us in His image, and that means clearly that He created spiritual beings, because He is spirit. And we also have to consider the fact that ethereal beings, including our souls, are a part of God. So we can honestly say that God created all life, because He *is* everything there is. The detailed facts about souls being part of God are even today hard to conceive for many people, especially when the mind is blocked by scientific or religious dogmas. (Gerry Spence said, "I would rather have a mind open by wonder, than one closed by belief.")

For the people who lived two thousand years and more years ago, it was impossible to conceive such facts. Therefore, the Bible and other Scriptures had to give a simplified version of the whole truth. We must not forget that at that time the earth was flat and at the center of the world in the minds of people. Things like solar systems, galaxies or universe were unknown! God was conceived of as a deity you can talk to, who talked to you, and performed miracles. Some even believed that God preferred some groups of "chosen people", and would help to destroy enemies in wars. This, of course, is plain nonsense. But even today many people, and some religious groups, believe in such absurdities.

How ethereal beings create life or design life forms is anybody's guess. But now, at the beginning of the twenty-first century, we can make an educated guess at least.

Because life in the cosmos has existed for eternity, most of the time it is not required to design and produce new life forms from scratch. Whatever a planet needs is carried in and "seeded" in the area where it is required, but new designs are still going on.

For these spiritual beings it is no problem to design, handle or change DNA in a similar way, as we would handle the construction of a computer. They work with protein molecules as we can with computer chips. This is easy to conceive if we bear in mind that ethereal entities (souls) are of the finest substance in the cosmos (remember Thoughtrons?). Our atoms are very large objects for them and easy to work with. All they have to do is to take a related DNA, apply their great knowledge and modify this DNA into another program; then they grow the first pair of animals, or the first plant from it. Some sources say that the ethereal realm has laboratories that are completely out of the range of our imagination and comprehension. I have no problem believing that.

One more word about the problem of God creating all things as against creation by a group of ethereal beings. Because the universal mind, God, is not a single entity but everything, God is in reality the creator with the help of an infinite number of co-creators. How can anyone believe that one personal God created the infinite number of things and life forms of the cosmos? What a job! No, He applies the same method of delegating a huge job to a multitude of entities, just as managers do in big corporation. And as is always required in such a situation, He set the rules – the cosmic laws.

What are the design criteria for physical life forms and what are their order of importance? With some logic we arrive at three main criteria.

Most important is the ability to reproduce, because these biological beings do not last forever. And the desire to do so must be preprogrammed into the DNA. As we all know, every animal and every plant has its own method of doing this.

Next in importance is the requirement that the new being is a positive addition to the environment, or that it will complete or restore the balance of nature. Also, the new plant or animal must be able to mutate if environmental conditions change. Without the ability to mutate (to adjust), a life form will not exist very long.

Third, I think, is independence from the spirit and the ability to be self-sustaining, which means as much autonomy as possible. The large animals can be on their own to a high degree but the smaller a creature is, the more guidance and control is required through a free spirit. And for the very small ones, even spiritual power supply is necessary.

One thing is common for all biological creations, plants and animals alike. They are all interconnected! Some of these interconnections appear to be very "strange". For example, Japanese scientists studied monkeys on one of the many small southern Japanese islands. They performed experiments to test the so-called formative causation, an effect of Carl Jung's collective unconsciousness theory. One scientist performed a task unfamiliar to the monkeys. He washed a sweet potato at the beach before eating it. He did so day after day, and the monkeys watched. Then, just as expected, one female monkey imitated this behavior and did the washing from then on. One after the other, the monkeys copied the washing, and then came the great phenomenon. When a seemingly critical number of about a hundred engaged in this "monkey business", the monkeys on all the surrounding islands also started washing their sweet potatoes, although there had been no contact whatsoever between the monkeys of the various islands!

I think that is scary, because the same effect also applies to us. It may explain in part the so-called waves. The sex wave of the sixties, or the latest crime waves, or the massacre waves among underdeveloped countries could be examples. On the other hand, if millions of people condemn a certain crime, the number of these crimes will go down, because the thoughts of these millions may stop many would-be-criminals from committing the crime.

Plants are also interconnected. In many places on earth, the forestry experts watched in awe at the strange behavior of trees. Tree-killing moths went from tree to tree and began to destroy the forest. But they did not get too far. The trees ahead of the moths began to produce a poison and had it ready in their sap when the moths arrived.

A few years ago I read about some interesting tests with plants and animals in *Fate* magazine that point to a spiritual connection. Four glass-enclosed boxes contained the same soil, the same plants, the same light and climatic conditions. In three boxes speakers played different kinds of music. The fourth box had no music (it was the control). The plants that enjoyed classical music performed a little bit better than the plants in the control box. The plants with country music grew much better, but the plants that got hard rock music withered away and eventually died.

A similar experiment with cows was described in *Newsweek* a few years ago. The result was different, because the cows have a different idea of good music than plants. The cows on country music gave 5 per cent more milk than normal. The classical music resulted in 15 per cent more milk! But the cows who were exposed to hard rock tried to break loose and get out. Because they could not get out, these cows produced 20 per cent less milk. (Because of this article many farmers play classical music for their cows with good results.)

I reason that hard rock, especially heavy metal, has very much to do with the crazy behavior of some kids today. And because it must always be extremely loud to be fully enjoyed, many have lost much of their hearing. In a special report of *20/20* in ABC, they found that many rock musicians, who are subject to loud rock on an almost daily basis, had lost their hearing permanently and are practically deaf.

All the previous examples should be enough to show that all biological life forms are definitely much more than just biological automates. Each plant and each creature is also spiritual to a great extent.

There is one more thing to say about how ethereal beings create third-dimensional creatures, because this other method at their disposal also creates and controls objects of the world of matter, or what we call dead matter. The cosmic mind is thought energy. Everything is made from it. Some scientists knew this fact very early, but the materialistic scientists of today forgot that. Eddington, Jeans and Whitehead, physicists at Cambridge University in the early 1930s, adhered to the tenet: "The stuff of the universe is mind-stuff."

Ethereal beings are pure thought energy with full power. They are able to create or move objects by thinking, combined with desire. This is important to know, because from time to time one of these celestial entities incarnates as a man on earth in order to help us to find our purpose in life; they are then known as masters. And they have some of their normal powers available, even when incarnated on earth. Jesus was one of them, and there were, and still are, many others, now and in the past. All of these masters were able to perform instant healings, and create or move objects with just the power of their minds. I have the suspicion that such masters were involved in the construction of the great pyramids, the moving of the monstrous stone blocks at Baalbek, and other great achievements we cannot explain or imitate.

## Energies and Forces

### GRAVITY AND LIGHT

The most important force, in my opinion, is gravity. Gravity and acceleration are the main items of Einstein's General Theory of Relativity. So far nobody knows what gravity is. It is the one thing that just does not let us look at its cards, but we are involved with it every second of our life. But before I present my own theory about gravity, let us see what we know so far about gravity and how it seems to work. It is also an introduction for the reader who is not familiar with the rules of this force.

Gravity seems to be a force of attraction between all matter.

It is a force of infinite range and is vital to the operation of the universe; however, it plays no part in the internal, subatomic structure of matter. It holds stars and planets together and keeps orbiting systems orbiting. The gravitation of the earth keeps our feet on the ground. It is called earth acceleration and as a value it is expressed as 9.82 m/s$^2$. This means that a falling object will fall 9.82 meters per second faster after each second. (Measured at the surface of the earth.) Earth's surface acceleration is 9.82, Mars has 3.76, Jupiter is 26, and Saturn's is 11.2 m/s$^2$, just to name a few. This so-called g-value at the surface of a body in space is a product of mass and size.

According to the applicable formulas, gravitation will get stronger if we get closer to the center of a planet, the so-called "center of gravity". Newton had to invent this imaginary center in order to make his laws of gravitation work, and that is what we learn in school. Then came Einstein and introduced a radically new concept of gravity. According to his theory of General Relativity, the mass of a body in space deforms (warps) the space-time continuum around it, and in turn, the warping of space-time causes this to happen. Confusing? Yes it is, because it is a paradox.

Now let us investigate this strange thing called gravity. A good example would be the relationship of the earth and the moon. According to our reference books we learn that the earth has 3.68 times the diameter, 49.3 times the volume, and 81.5 times the mass of the moon. But the surface gravitation of the moon is not 1/81 part of the earth's gravitation, as should be expected, but one-sixth. Why? We learn that the moon's surface is at a ratio of 3.68:1 closer to the center of gravity (the center of the moon), and that the gravitation decreases with the square of the distance. Therefore, we have the diameter ratio $3.68^2 \times 0.01227$ (1/81.5), or $13.54 \times 0.01227 = 0.166$, which is one-sixth of the earth's gravitation ($=1$). This must be right, because everybody knows that the moon's surface gravitation is one-sixth of the earth's.

If we follow these calculations, we find that the mysterious "center of gravity" is a mathematical requirement in order to make Newton's laws work. It is only an imaginary factor. Newton had to introduce this center of gravity, or nothing would work. Something is wrong with our picture of gravity if we have to imagine a super-atom of infinite mass in the center of a body in space.

But even though we know that the center of gravity is only a mathematical reality, and does not really exist, some scientists went down to the lowest point of a two-thousand-meter-deep gold mine in Johannesburg in South Africa. And they got their "proof". Down there gravitation measured just a little bit stronger than at the surface. But what really happened was that down there they were two thousand meters closer to the large core of heavy metals, which provides a high percentage of the earth's mass. The maximum gravitational value is not at the surface, but at an optimal depth of a few miles because of the large core. So what is wrong? Why do we need such an imaginary factor within our formulas? We will see.

Let us take another look at gravity as science sees it today. As unbelievable as it sounds, it is really possible to develop physical laws and formulas without knowing what the subject matter really is. We do it all the time! Isaac Newton's laws of motion and gravitation are still fundamental to celestial mechanics today, but he did not have any idea of *what* gravity really is. Now we have electrical and electronic engineering, but nobody knows *what* electricity actually is. Ballistic and other related disciplines in physics work with kinetic energy, but we have no concept of this magic energy; it is beyond visualization and comprehension. In mathematics we work very well indeed with values of even less than nothing and with infinity.

When Albert Einstein developed his theory of General Relativity (1915), he confronted the same situation. He did not know *what* the three main elements of his theory (space, time and gravity) actually are, but in ten years of hard work he produced a system of laws and equations that seem to work. But what serves as the basics of his ideas are, to say the least, strange. Let us take a closer look:

1. The gravitation of a body (mass) in space changes the geometry of space-time around this body, causing it to become curved.
2. It is this curvature, or warping, that causes the motion of bodies (causes gravitation).
3. Thus matter tells space-time how to curve, and space-time tells matter how to move (have gravity). This is repeated ad infinitum.

If we think about this and apply logic, we will be reminded of a dog, chasing his own tail, or a man, pulling himself out of the water with his own hair. But strange as it sounds, we can read this in every book on the subject, even in the official works of reference. Everybody accepts this paradox. The only reason that I can see for this acceptance is because Einstein said so. I cannot concur because it defies logic.

Einstein never explained *how* a mass changes space-time and, of course, he could not explain *how* space-time causes a mass to develop gravity, needed to change space-time itself.

Before I present my own theory of the nature of gravity, I shall show you the reasoning that led to the development of my theory.

First, I have my own questions on this subject. These questions and the above paradoxes triggered this theory.

1. Why do we need that imaginary center of gravity in our formulas of gravitation?
2. How do the atoms of a mass in space manage to pull through millions of light years of space (or curve space-time *a la* Einstein)? For example, how does every single atom of our sun reach out to Pluto? They are involved because they are each a part of the sun's mass.
3. What is gravity in detail, and how does gravitation work?
4. What is the strong force that holds the nucleus of an atom

together? How do the gluons carry a pulling force from one particle to another, resulting in the mysterious strong force?
5. How can the *pulling* strong force of extremely short range ($10^{-13}$ cm) *push* out for miles at nuclear fission (as with the atom bomb)?
6. Why is the released energy from a single atom at nuclear fission equal to the $c^2$ of $E=mc^2$?

Of course, the last three questions are for quantum mechanics and not for relativity, but in my theory both belong together. The answers to the above questions are the basis for my theory of gravity, solving the paradoxes of General Relativity and more. It requires a *complete U-turn*, the only way to get out of a dead-end street; and cosmology is in a dead-end street! Therefore, my theory turns the whole thing 180° around, completely upside down, and makes gravity a *pushing force*.

The problem with the current concept of gravity is the illusion that it is a pulling force. Who can imagine something pulling through millions of light years of empty space? What is even worse is that some disciples of quantum mechanics hypothesize a gravity-causing particle, the "graviton". That must be quite a subatomic particle if it can pull through light years of space without a rope or chain. This basic concept of gravity being a pulling force is just plain wrong.

Some readers will now point to a magnet and say that we can watch the performance of pulling through space when it catches a nail. It is the same kind of illusion that we have with gravity!

I will show you what I think about a magnet, and why it leads to the only solution for gravity that makes sense. So let us reason about a magnet in a new way.

Without proof, a theory remains just that; a theory. Besides the so-called solid proof in a laboratory, or through definite observations, scientists also accept proof based on a "thought experiment" of logical reasoning. They accepted Einstein's theories which are mainly based on thought experiments. The following reasoning is something I did not find in any book or article, and I wonder why.

The idea is that a magnet poses a problem similar to the gravitation of a body in space. Let us say that we have a very strong magnet, capable of holding one hundred pounds. It can easily lift one pound of steel, so it jumps up a half inch through the air. This needs energy! And our magnet can do this millions of times, and that accumulates into a lot of energy!

Where does the energy come from? This permanent magnet has no detectable energy source. The weight remains constant, so it does not change mass into energy. Heat transformation does not occur either. And we did not apply any other source of energy, like electricity. It is just there and works without any energy consumption. It is a miracle, the perfect power source, generating a force out of nothing. But by the laws of physics, there must be a power source!

Logical reasoning leads us to the only possible answer. We must postulate that magnetic energy exists everywhere, in every spot of the universe, arriving at each infinite small point from all directions. It is polarized and equally strong from all sides. Therefore, every point is in perfect balance and nothing happens. Except when it is disturbed in some way; and our magnet is one way to disturb this balance, one way out of many.

Just by the way its atoms and molecules are organized and oriented, it prefers one polarity and its direction, and so liberates the magnetic energy that is already there, *without the use of any own energy*! (It has none anyway.)

In a way it is similar to the nozzle at the end of a garden hose. The nozzle has no energy of its own; it is only built in a certain way. But it can change the water which is under pressure behind it, into a powerful jet of water, without the use of any energy of its own.

If we send an electric current through a wire coil, the empty space inside the coil will become a magnet. This effect is called electromagnetism. But here is a great, worldwide misconception: It is believed that the electricity transforms into, or produces the magnetic force. This is not true. It sets free a magnetic energy which is already there, doing in effect the same what a permanent magnet does!

Now, after getting a grip at the principles and the logic behind such "miracles", we can approach the problem of gravity, where the same principle of reasoning applies: Without any use of an own, built-in energy, a body in space can get an already existing force to do the work.

Out there, between Mars and Jupiter, is the asteroid Ida with a diameter of about 35 miles. It is an absolute dead chunk of rock, as dead as it can be. Ida has a tiny moon, about one mile in diameter. For the last few millions, maybe billions, of years Ida forces its moon into orbit. This requires very much energy over all that time.

Here we have the same question as before. Where does that energy come from? There is no detectable source for the energy in the asteroid. We cannot even think of any. But physics requires a source of energy.

Again, logical reasoning directs us to one explanation. We must postulate the existence of gravitational energy at every point in the universe, arriving from all directions with equal force. In this example, Ida influences this omnipresent energy in such a way that the moon is forced into orbit. How, we will see when I explain my theory of gravity.

The above logical reasoning shows that all of us, including Newton, Einstein, and all the other scientists, took an obvious miracle for granted, without wondering how a magnet or an object in space can generate an energy form (or force) out of absolutely nothing. Newton and Einstein produced laws and equations that work, but both did not know what gravity really is. Einstein tried with his strange and exotic idea of a curved space-time that results in gravity. He never could explain how that works. For me it is just a strange paradox.

Magnetism and gravity are not like motor ships with built-in power, but rather like sailing ships that take advantage of an energy form that is already there. We just have to give credit to the brilliant design of a causal cosmic intelligence, that is all!

And here, before I begin with my own theory of gravity, I would like to repeat the quote at the beginning of this chapter. The well-known physicist, John Archibald Wheeler (b. 1911), once remarked on the possibility of a unified theory of everything (TOE):

> When the solution to the great picture comes, it will be so simple, so pure and so obvious, that we will marvel at ourselves for not having seen it all along.

Maybe I have found it. It is so suspiciously simple and obvious – and it is logical. (I hope he will be able to read this book.)

*My different look at gravity*

We postulated that gravity is a pushing force. Because the cosmos is an *infinite entity*, we can postulate that a gravitational force arrives from all directions, and its rays go into all directions. Let us assume it is this way and we will call them "gravity rays". Also, let us postulate that these rays are of such a high frequency that the hardest gamma rays are long-wave monsters in comparison. And one more assumption: The diameter of these rays is – you guessed it – infinitely small. These rays are the basic *active* fabric of the cosmos, and not passive like air or the famous "ether". And they contain energy! They zip through a planet almost without resistance, but while doing this they transfer an almost infinitely small amount of kinetic energy to every atom of the body in space. In fact, they transfer this impulse to every one of the smallest subatomic particles.

If the body is alone in space, the effect is completely neutralized because it works equally from all directions. But the rays that exit from an atom or a large body in space are weaker than at entry, in direct proportion to size and density. Now let us think about one detail. The rays that pass from the surface to the center are much stronger than the rays that pass from the center out to the other side. The result is that the rays force and holds the body together.

But nothing is alone in the universe. Therefore, in our example of earth and moon, a *shielding effect* occurs. The gravity rays that hit the moon from the direction of the earth are weaker than the rays from all the other directions, because the earth has absorbed some of the energy of the rays that passed through it. As a result, the moon is pushed towards the earth. And the earth is subject to the same effect through the moon, but to a lesser degree. The strength of gravitation and the degree of shielding depends on the size and density of the body in space. The effect on a single atom is very minute, but the sum total of all the atoms of a planet results in a strong force.

[look out for description of the human brain]

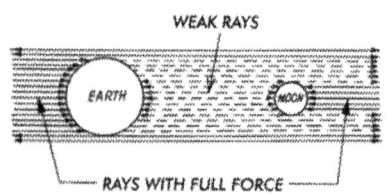

**The shielding at earth and moon (Principle)**
(ONLY ONE DIRECTION SHOWN FOR CLARITY)

FIGURE III

Now I would like to repeat the earth-moon example under this new 180° upside down aspect. We know already that the earth's diameter is 3.68 times the diameter of the moon. The density is 1.65 times greater than the moon's. Now, if we multiply the ratios of the diameters and the densities, we have 3.68 x 1.65 = 6.07 times the gravity of the moon. Is not that much nicer? As we can see, diameter and density are all we need to calculate the gravity of bodies in space and their relation to each other. Shielding is the name of the game! The earth shields the moon and the moon receives less push from the direction of the earth than from all other directions, and so the moon is pushed towards the earth. (See Figure 3)

The beauty of this new picture of gravity is that Newton's laws and formulas are still fully valid, but we have got rid of that imaginary center of gravity. And of course, we have to change our way of thinking about gravity. It is a U-turn!

Instead of an attracting pulling force that mysteriously works through empty space, we now have the following theory: Each body in space would be in perfect balance if alone in space, because the gravity rays push from all sides with exactly the same force. However, this is never the case in the cosmos. A moon shields a planet a little; the planet shields the moon more. The result is a higher pushing force from outside than from between. The two are pushed together. And so it goes on and on through the cosmos. With this model we have no problem imagining that this effect can reach out into space over millions of light years – in fact all the way through the universe.

The new picture of gravity in the cosmos is now this: The system of gravity rays from and into all directions is the basic fabric of cosmic space. The shielding effect of every mass disturbs this fabric by weakening the gravity rays, resulting in the gravitation for a shielded object. Let us take the earth. A moving object, a meteor for example, that has a straight course towards the earth, will be accelerated through the shielding effect until it hits. If the course is such that it bypasses the earth, the shielding effect will bend the trajectory towards the earth and it will move in a different direction. But if distance and speed are just right, it will be forced into orbit. And, in the case of a star, a beam of light passing by at a close distance will also bend due to the shielding effect of the star.

That is how the area around a body in space is disturbed according to my theory. Einstein's space-time consists of the three geometrical dimensions plus the time. I cannot image how Einstein could envision that this space-time condition could be curved, and then cause the gravitation of a mass. The fact that he created equations to support this paradox is a great accomplishment, I think. But if we look closely at it, we see that my "disturbance" is in effect the same as his, but the difference is that it can be conceived.

There is still much more to my theory of gravity. But before I continue I must direct a few words to the professional physicist who may read this. One mechanism of gravity was thought of a few times in the past, the first as early as 1750. It was similar to mine, but not the same. The idea was that many particles moving in space at a very high speed in all directions are only slightly absorbed when zipping through matter. Because they are coming from all sides, all the impulses balance on a body in space. But in case of the earth, for example, there is the sun on one side, and *fewer particles* coming from the sun than from the other side. Therefore, the earth gets an impulse towards the sun. All known laws of gravity are still valid.

Then scientists discovered that this principle cannot work for one important reason. The great American physicist Richard Feynman (1918–88) explained this impossibility best, using the earth as an example. While orbiting around the sun, the earth would impinge on *more particles* on its forward side, exactly like when running in the rain, you get more in your face than at the back of your head. This causes a resistance to motion, slowing the earth down until it stops. Nothing can keep moving for long in the universe! Conclusion? This mechanism for gravitation is impossible. And if a man like Richard Feynman said so, it must be true. Right? *Yes and no!*

In the above discussion I have stressed the words fewer particles and more particles. They represent the reason why the above counter evidence does not apply to my theory, but only to the wrong ones Feynman talked about. My theory neither talks about particles, nor about a quantity.

Not particles, but energy-containing rays of infinitely narrow width give a minute impulse to *every atom* of the earth. Because the network of the gravity rays is of infinite density, not a single atom can escape! This means that the earth can never impinge on more rays. In order to be subject to more ray impulses, more atoms are required. Everybody can see that moving will not result in creating more atoms in the earth. Moving has no effect on the forces of gravitation. And, of course, not fewer, but weaker rays arrive from the sun.

So much for the people, especially physicists, who learned that gravity as a pushing force is not possible. Sometimes you just cannot see the forest for the trees; even the brilliant Richard Feynman did not see it.

Because the shielding effect (gravity) gets weaker with the square of the distance – and distances in space are very great – gravity appears to be the weakest force. Very weak! But in reality, gravity rays are the strongest force of the cosmos, only a bit weaker than the causal force of the universal mind. Gravity can be strong beyond imagination if *completely* shielded from one side by a black hole, for example. Some physicists may not believe this, but maybe lifting a 94-pound bag of cement may change their mind.

But before I get to the big guys in space, I must talk about the smallest.

It is a well-known fact that any surface gets larger in ratio to the volume when the object gets smaller. If you do not know this fact, think of it this way: The total value of gravitational force on a body from all sides (what holds it together) is in direct proportion to the diameter and density, but not to the volume. What this means is that a sphere has a ratio between surface and volume. This ratio changes with the size. Let us assume a sphere has a surface of $24^2$, then the volume is $8^3$, a ratio of 3:1. At half the diameter, the surface will be $6^2$ and the volume $1^3$, a ratio of 6:1. And at a quarter of the diameter (half again) the surface will be $1,5^2$ and the volume $0.125^3$, a ratio of 12:1. The smaller a body gets, the greater the ratio between surface and volume.

I will illustrate this with two easy examples. Big animals have problems with heat dissipation because the surface of their bodies is relatively small in ratio to the great volume of their bodies. They sleep stretched out. (And elephants must circulate all their blood through their large ears for cooling.) Small animals have a large body surface in ratio to their body volume. Their problem is staying warm. They sleep curled up.

We see that the ratio doubles each time we cut the diameter of a sphere in half. I have to illustrate to what extremes such a mathematical function leads, so you will get the right idea about the reasoning in the following paragraphs. Imagine that you could fold a sheet of paper, 0.1 millimeter thick, fifty times. Of course, it is not possible, but imagine. As you know, each time you fold, you will double the thickness of what you got at the previous fold. How thick would you guess the final pack to be after fifty folds? Believe it or not it will be more than the distance from the earth to the moon, much more! In fact, after forty-two folds we will already have 440,000 kilometers, way past the moon. After fifty folds we will have 113 million kilometers – that is from here past the orbit of mercury – close to the sun.

Now let us go back to our favorite earth-moon example. The moon has one-sixth of the earth's gravity; but if we compare the ratio of their masses, 81:1, then the gravity of the moon is 13.5 times larger than the earth's, relative to their masses. And to top this, the density of the moon is only 0.606 of earth's. My point is that the smaller a body, the greater the gravitational force that works on it, relative to the volume!

Next we will have to recognize that these gravity rays from all directions are of an infinitely narrow width, and therefore work on each subatomic particle. The effect on a planet is only the sum total of the effects on all the subatomic particles of a planet. We also have to recognize that *for the gravity rays the nucleus of an atom appears to be a large body in space*!

Now, if we imagine the doubling game we just played in reverse, and do this in millions of steps down to the diameter of a nucleus, then we arrive at a more than astronomical ratio of surface versus volume. The gravitational pressure in these subatomic realms is almost infinitely high in ratio to the volume. It is very strong at the level of a molecule, but a force beyond imagination at the nucleus of an atom. Under these conditions the gravity rays would hold a nucleus together even if it would be composed only of positive-charged particles; it is that powerful.

This is the *strong force* that pushes the nucleus together from all sides, and not the theorized pulling force with an almost infinite short range of $10^{-13}$ centimeter. In my theory, gravity is the primary force in the macrocosm *and* the microcosm!

What keeps the nucleus from collapsing? It is the causal cosmic energy, spinning at the speed of light. In fact it is the causal energy of the cosmic intelligence, which caused the creation of the cosmos. (It is also capable of changing location at superluminal speed – instantly!) Because the velocity of light enters the formula for the centrifugal force as a squared multiplier, and the dividing radius is close to nothing, the resulting counterforce from the inside of the nucleus is also beyond imagination and keeps a fine balance. As a nucleus gets larger from element to element, the outside and the inside forces get weaker. After 92 such steps from the hydrogen atom, the condition for balance is over. Everything beyond element 92 (Uranium) is unstable and can be created only artificially. (This is also an explanation for the fact that some of the heavy elements are radioactive. They are at the limit of balance.)

Let us go back to the spinning force from the inside. The $c^2$ at an extremely short radius is the force that gets free if the nucleus is cracked by any means. *It is the $c^2$ of $E=mc^2$!* The result is nuclear fission. This atomic explosion with the power of $c^2$ is possible because this force from the inside has the right direction in the first place – out. It is impossible for the now theorized strong force, pulling towards the inside with a short range close to nothing. (Mr. Spook would say, "It is not logical.") An explosion over a long range of many miles can only be caused by a pushing force from the inside.

Now some alert readers will ask, "What about the atoms at the surface of a planet? These atoms get full force from the outside, but weaker gravity rays from through the planet. How can they remain intact?"

The answer to this question is how gravity works at the subatomic level according to my theory. It is an answer to an eternal quest. A planet is no great barrier for the gravity rays. They zap through with very little loss of energy. The forces that work from the outside on the nucleus of an atom are of such gigantic proportion that the small difference has negligible significance. It can be completely neglected. But because there *is* an almost infinitely small difference, the atom is pushed towards the center of the planet (shielded by the planet). If we look at a solid object at the surface of a planet, we see that all the atoms of the object are subject to this effect, and as a result the whole object is pushed towards the planet. The number and the size of the atoms determine what we call the weight of the object. Each atom follows the same rules as the sum total of a multitude of atoms in the form of a person, moon, planet, star or galaxy. And, of course, it is the same the other way around; every body in space follows the same rules as its atoms.

To summarize, each object is pushed towards the earth, because it is shielded by the earth. But in reality, each atom is shielded and the effect of gravitation is the result of the sum total of all atoms of such object (its weight).

Like everything else in the world, this basic principle of gravity being the strong force also has a limit. The limit is a very large body in space, our sun for example. The huge diameter of 61.39 million kilometers and the relative high density weakens the gravity rays too much for the atoms in the outer shell of the sun's body. They break apart and we have nuclear fission. I think that this effect could be an explanation for the extreme high temperature (2 million °C) of the sun's corona, which the surface has only 8,000°C. Deeper in the sun, where high pressure adds to the temperature, nuclear fusion from hydrogen into helium takes place. The helium boils up to the surface and causes the strange look and the flares we see.

Some stars reach a size that is too big. Now almost all gravity rays are absorbed and the giant collapses under the tremendous force of gravity from all sides. While collapsing, the density goes up and so does the surface-to-volume ratio, which makes the situation even worse. (Remember the effects on an atom: The smaller, the stronger.)

At the end, a relative small body is so dense that it absorbs *all* of the gravity rays. Every object that comes too close to this super-concentrated body will be pushed into it with the maximal possible force, hitting it at the speed of light. Nothing can escape; even light has no chance to get away from this body. (Why and how we will see when we arrive at my theory about the speed of light.) Because even light cannot escape, we cannot see this super-compacted star anymore, not at any wavelength. Therefore, all we can see is a black spot in the picture of space. This fact gave somebody the idea to call such an object a black hole. (We know it is not a hole, just the opposite.)

Lately, some scientists have come up with some very exotic ideas about black holes. If you read them and apply common sense, they are definitely funny. In the last few decades scientific journals and magazines have carried dozens of different theories about black holes. Even serious books contain these tales. The following few examples should be sufficient, even for a non-scientist, to show that we just cannot take them seriously.

The very best, I think, is the so-called singularity in the center of a black hole. Einstein started to use this word for a singular entity. Now it is being implied that objects pulled into this singularity will be squeezed out of existence, and because all laws of physics break down within a black hole, this now non-existing object will be spilled out at the other side into another universe. Wow! If a black hole rotates, these scientists say, its singularity is a tiny ring in the center of the black hole. Objects that pass through this ring emerge on the other side at another time and in another spot of the universe (through the Einstein-Rosen bridge). It looks like all we need for a time machine is a nice, cute black hole.

I guess that the inventors of these tales have watched too much *Star Trek*. I do not understand these ideas about black holes, and I am sure that Newton would like my definition better. And I guess that the great French mathematician and astronomer Pierre Simon de La Place (1749–1827) would certainly smile, because he was the first person who theorized the existence of black holes in 1795. Then came *Star Trek*.

One question remains. What is the fate of a black hole? It is a big problem because a black hole can increase in size, but never explode. Will the whole cosmos end up in nothing but black holes? Some scientists believe so. These theorists know that a black hole cannot explode, but they believe in the Big Bang, which is in fact the explosion of the ultimate black hole, containing the mass of a whole universe. But according to my theory of the nature of the universe, such fate is nothing to worry about. Each black hole will eventually reach the outer rim of the universe. There it will leave our universe for a long journey through the "inter-universal" cosmic space, or it will meet an incoming antimatter black hole from another antimatter universe and both will annihilate each other in the form of a quasar. This is the only way to destroy a black hole, and it serves as a way to replenish the universe with new particles for matter. The quasars send their gamma ray photons and other radiation into all directions, providing the raw material for new worlds. I have already explained why everything moves out to the edge of the universe, gaining on mass and velocity being the reason. And black holes gain mass very fast.

Permanent, steady addition of kinetic energy to a body results in *acceleration*, which means going faster and faster with every second. The effect of acceleration on a body is exactly the same as the effect of gravity, but they are not the same! Because of the identical effect, acceleration and gravity can add, subtract, or cancel each other out for balance, as in a stable orbit for example.

What is *inertia*, the tendency of matter to remain at rest and resist an outside force that wants to move it? What causes it? So far we have no theory for the mechanics behind this strange behavior of a mass. (Inertia is also the tendency of matter to keep moving if it moves.) I think my theory of gravity can come to the rescue and provide an explanation.

First a definition is needed. If a car stands on a pavement and we want to push it, a strange effect takes place. At the start we have to push very hard for a while before it begins to move. But then, as soon as it moves, we need very little force to keep it moving. That is inertia, the resistance of an object to be moved. In the case of our example we have the impression that a few people try to hold the car in its position, and we have to overcome their holding power.

Replace the holding people with the gravity rays, and we are on the way to an explanation. These rays not only hold a body together, but they also hold it in place. Depending on the mass (number and size of atoms), we need more or less force to overcome the holding power of the gravity rays and get the object moving. Or in the language of physics, to overcome inertia.

In physics and astrophysics, as well as in Einstein's Theory of Relativity, there exists a so-called "body at rest" concept. It is very famous. A little reasoning tells us that such a thing does not exist anywhere in the universe. A "resting" rock on the surface of the earth is not at rest at all. That rock is moving along with the rotation of the earth, and by the orbit of the earth around the sun. The rotation of our sun around the galaxy, and the rotation of our galaxy around the center of the universe, 450 kilometers per second, must also be accounted for. If we see our rock from this viewpoint, we have logical proof that gravitation forces a body in space to keep on doing whatever it does – remain in place, keep on moving, or accelerate if shielded. This is my definition of inertia.

Now we can see how another unexplained energy works – the strange *kinetic energy*. It is a force that is transported along in a body of mass by the gravity-caused effect of inertia. For example, the explosion behind a bullet in a gun uses most of its energy to overcome the inertia of the bullet and then accelerate it. This does not change the bullet in any way. The size and weight remain exactly the same, but the mysterious force of kinetic energy is now lurking inside. How? The effect of inertia, caused by gravity, keeps the bullet going along at that speed. At the moment it hits some target, all the energy that was needed to overcome its inertia is now released in the form of sudden deceleration. And because this occurs in an instant, all the energy is squeezed into this extremely short span of time and can do so much damage.

In addition to the above forces we have the many so-called *vector* quantities. The best known is the centrifugal force of a rotating mass. This vector force is the most important one, because it keeps everything in the cosmos moving and in balance. If you are not familiar with the term "vector", I have a simple, well-known example. The leaning tower of Pisa has a mass-center point located somewhere in the middle of its mass. From this theoretical center point, a gravity vector points vertically down and ends up within the foundation area. The tower still stands. At the instant this vector leaves the foundation area, because the tower leans over more than it should, the tower will become history.

One last word about the atom. How does the causal energy perform counteracting from the inside? Of course, I do not know nobody does. However, I do have a theory which I mentioned shortly before. The whole cosmos and our universe is powered and under control by rotation and orbiting. So why should it not be the same inside the atom? We know that there is no reason for an exception to this rule in the form of the Big Bang explosion. So why should there be an exception that is pulling towards the inside of an atom in the form of the so-called strong force? No, only a spinning energy can counteract the tremendous force of gravity from the outside of the atom. If we consider the very short radius and the high rotating velocity (speed of light), and apply them to the formula for centrifugal force, the result is a force beyond imagination. That is the required counterforce. And what causes the rotation in the first place? The causal cosmic mind energy.

Please remember that this rotating velocity, the speed of light (c), enters the formula squared. The $c^2$ of Einstein's $E=mc^2$ is at work here, pushing to the outside. That is why mass releases so much energy when the atom is broken and the $c^2$ gets free. I guess that Einstein had similar thoughts when he came up with his $E=mc^2$ – but I also believe that his thoughts were in the form of mathematics. He could not have my clear view, because for him gravity was a force of attraction,

I know very well what I am doing here. I have replaced the three basic forces at the atom, the strong, the weak, and gravity with just one, gravity! And the new theory is in accordance with many religious scriptures of the world, and the words of the masters. In one of his readings, Edgar Cayce said that the macro-cosmos and the micro-cosmos work in the same way – as in the universe, so in the atom!

What did I do differently? I included the causal energy, the power of God, and it worked! This is what our materialistic-oriented scientists reject and do not consider. And just for this reason, they cannot arrive at complete solutions. It is the same for physics what instinct is for biology.

After all of this, the informed reader may say that the General Theory of Relativity had proof beyond any reasonable doubt when Einstein could solve the problem of the perihelion of Mercury with his formulas. The problem was this: Mercury orbits around the sun in an elliptical trajectory. Observation showed that the ellipse of Mercury's orbit also rotates very slowly. Nobody had an explanation until Einstein explained and solved it. All I can find in the reference literature is that his calculations came very close to the observed value (no data).

I tried the same according to my theory. I started with the fact that the gravity rays need 4.64 seconds to pass through the sun. The sun makes one revolution in 26 days. Then I calculated the offset of the shielding from the centerline of the sun, caused by the 4.64 seconds and the rotation of the sun. Then I calculated the effect on Mercury which has a year of 88 days. My final result was a perihelion advance of 42.7 arc seconds per century. The observed value is 43 arc seconds per century. As you can see, my theory works too. And I did it without the fancy tensor calculus mathematics of General Relativity.

To finish the topic of forces, here are some thoughts about the problem of mastering gravity. (Listen now, UFO fellows.)

When Michael Faraday (1791–1867) proved in 1845 that magnetism effects light, and when Einstein's theory that gravity can bend light was proven in experiments, then another possibility arose – the reverse should be true, according to the laws of physics. Electromagnetism should be able to effect gravity rays! So let us imagine that we have the technology to do just that.

If we are in a spaceship and our power system is off, nothing happens because gravity rays push equally from all sides. Now we turn our system on and split some of the rays on the side we want to go. The split rays will bypass the ship, and as a result the full gravity from behind will push the ship into the void in front. The ship will not just move, it will accelerate in a free fall into the wanted direction. Because the whole ship is in a free fall and the crew members are part of its mass, they will not feel anything, not even at accelerations of hundreds of *gs*.

To hover in the air, only a very small amount of the rays above have to be split (just enough to cancel out the gravitation of the planet). Then, if in danger or to show off, we turn on the full power, the ship will shoot up into space. The crew will feel nothing because it is a free fall up! This would be the better-than-rockets technology to develop for trips within our solar system, as I mentioned in a previous chapter.

If our scientist would only accept my theory and do research in that direction, we could have the following conditions on earth: All airplanes would fly without wings, engines or fuel by the same principle as spaceships. Thousands of designs are possible to generate electricity for cars, houses, factories, ships, etc. For example, to generate electricity for a house we can connect a generator to a very heavy wheel. If we reduce the gravity rays (from above) for only half of that wheel, the wheel will rotate and turn the generator. This is not science fiction! If my theory is right, and I am sure it is, then this will be the clean power source of the future, without fuel and absolutely free. And this power source never runs out, because gravity and electromagnetism are always available everywhere in the universe, free for us to use.

And what will the maximum speed be for such spaceships? It is the speed of light, because the gravity rays cannot push anything faster than their own speed, 300,000 kilometers per second. This is why no object can exceed the speed of light! It is not because the object reaches at that speed an infinite large mass, as Einstein wants us to believe. Where does that imaginary mass come from anyway? I think it is as simple as everyday life. If my car has a top speed of 60 miles per hour, I cannot provide a push to another car at more than these 60 miles per hour. And the gravity rays are in the same boat.

*Speed of Light*

While the problem of gravity was relatively easy to solve, requiring only a U-turn in our way of thinking, it is much harder to find a solution for the "worst" of all energies, the electromagnetic radiations. Our arbitrary classification system calls them radio waves, infrared, visible light, ultraviolet, X-rays and gamma rays. They are different by their frequencies, low for radio waves and very high for gamma rays. But all of them have one thing in common. They move at the same speed, the speed of light, 299,792.46 kilometers per second in a vacuum. And that speed has no tolerance; it is ± zero, (Generally, it is referred to as 300,000 kilometers per second.)

But what makes light a real problem for physics is the fact that it violates the classical concepts of relative motion. The problem is the so-called "constancy of the velocity of light". This principle conflicts violently with common sense, because it disregards the transformation laws of physics. Why?

Light has always the same speed, under all possible conditions and under all possible frames of reference. And this should be impossible according to classical physics. Let me explain what that means with an example. A meteor approaches the earth at a speed of 30,000 kilometers per second. According to the law of relative motion (Galileo) we should measure the speed of the light from this approaching object as follows: 300,000 kilometers per second from the speed of light, plus the 30,000 kilometers per second from the meteor which makes it 330,000 kilometers per second. But that is not the case; we measure exactly 300,000 kilometers per second! The speed of light is always totally independent of the speed and direction of the source or the observer. And that was a great headache at the end of the last century.

When Albert Einstein learned about these facts, he saw the mission of his life in these problems and went to work. In

1905, he published his new theory of Special Relativity, which solved, in his opinion, all the problems of the constancy of the speed of light. Everyone laughed about it in 1905, but today it is almost religion and "proven" because his equations seem to work.

What did he do? He took Galileo's laws of relative motion and improved it by adding the electromagnetic radiation. It required hard mathematical work, because it was somehow similar to forcing a square bolt into a round hole.

What I wonder about is that no one seems to see anything wrong with some of the consequences of this theory. In 1905, when he tried to explain his theory with the help of examples, he used the best model available at that time for his so-called "thought experiments"; in this case a train and lightning. Today we have much better versions that give the same results, but are better to understand.

One is a moving room inside a rocket that is moving with a uniform velocity. In the exact center of this room is a flashing light bulb. At the precise moment that the center of this room passes by an outside observer, the light flashes. The observer knows and sees that the room is moving. He observes that the light reaches the rearward wall sooner than the forward wall, because both walls are moving during the time the light speeds through the room.

But an observer inside the room observes the same event differently. He does not know that the room is moving, but he knows that light has always the same speed. He observes that the light strikes both walls simultaneously. While strange, according to his own writings about Special Relativity it is true that we can have these two different truths for the same event.

I do not like that, because it is not logical. A closer observation of this thought experiment shows that he drew the wrong conclusion. Here is what I think: The outside observer observed the true condition. The inside observer, being ignorant of the fact that the room is moving, *has to wonder why* the light hits the rearward wall first! That is the right conclusion.

If we substitute the room, light bulb, and the two observers with two planets, millions of light years apart, and if we put a sun exactly halfway between them, and then let all three move very fast along in the same direction, we have the same set-up in principle. What happens here? The people on the front planet observe a red shift of the sun's light spectrum, and the observers on the rear planet wonder about a blue shift. What is the reason? The front planet speeds ahead of the sun's light, and so the frequency is stretched (Doppler effect). For the rear planet it is the opposite; the frequency is compressed (higher). Everything can be true in only one way!

Another consequence of his theory is something we find in every reference book and is, to my opinion, even stranger than the thought experiment just explained. Here is what everybody believes because Einstein said it:

1. With increasing velocity, an object gets shorter until, at the speed of light, the length becomes zero (it disappears).
2. With increasing velocity, the mass of an object increases until, at the speed of light, it becomes infinitely large.
3. With increasing velocity, the time for an object slows down until, at the speed of light, time stands still.

This is really surprising to say the least! If we apply logic to these three "facts", we have to confront the following paradoxes:

1. At the exact instant an object reaches the speed of light, it disappears *and* will have an infinite large mass.
2. When the above paradox occurs, the speed of light must do without a time component, which is a mandatory part of the term "speed", in this case kilometers per second. No speed at the speed of light anymore! This is not even funny.

You will certainly have to question why nobody else saw these shortcomings of Special Relativity. I am very sure that many scientists, if not all, saw them too. (Maybe some are too impressed by the fancy mathematics.) I have my personal proof for that in the form of a question. Why are scientists still trying to find ways to test Einstein's theories after almost one hundred years? Because these theories are so abstract and exotic that testing them is almost impossible. They deal with conditions that we cannot replicate in a laboratory. And there is a big difference between all scientists and me. I have nothing to lose – no job, no government grant, no reputation in any danger. This means that many scientists may think the same way as I do, but they cannot say it.

I know that Einstein was an excellent physicist. He was also a very good lecturer. But I believe that he made a mistake, when he got the idea to "improve" the law of relative motion by adding the electromagnetic radiation around 1900. Light and gravity do not belong in relativity! Both have to follow their own set of laws. The law of relative motion is for the behavior of matter, and Einstein should have left it alone. Because of this mistake at the very beginning, he had to draw all those illogical conclusions in order to get light squeezed into relativity.

Back to the speed of light. What can we do to get around these paradoxes? Accept them still as a fact after all examinations? As you already know, I cannot. And I know that Einstein understands that I have to replace his theories with others. I am sure about that, because he had this to say in his book *The Evolution of Physics*:

> There are no eternal theories in science. It always happens that some of the facts predicted by a theory are disproved by experiment. Every theory has its period of gradual development and triumph, after which it may experience a rapid decline.

I am sure that he understood that these words also apply to his own theories. And I have his permission to develop a new view, because he wrote in the same book:

> Successful revolt against the accepted view results in unexpected

and completely different developments, becoming the source of new philosophical aspects.

Before I begin with my own theory about the nature of light, I would like to add a few questions of my own. Nobody else seems to have asked them. I believe that these questions are essential for approaching the problem from the right viewpoint.

1. How does a tiny photon without mass, and therefore without any kinetic energy, manage to travel through billions of light years for billions of years? What is the power source?
2. Why and how does this photon keep its speed constant at exactly 299,792.46 kilometers per second, ±zero? And it keeps this speed constant even while passing through strong gravity and magnetic fields. There must be a speed regulator, but how?
3. How do all frequencies of electromagnetic radiation, from gamma rays down to radio waves, behave in exactly the same way with regard to speed?
4. How can gravity bend a beam of light, if photons have no mass for the gravity to act on? Some say it is only energy, but $E=mc^2$ tells us that energy equals mass.

And now I will explain my theory of light propagation. To find the right solution, I set myself the following rule: *first*, we must find out *why* and *how* the speed of light is always constant. *Then*, we can go from there, having common sense on our side again. (That is what Einstein did not do.)

Besides the constant speed problem, there is another problem with light, even older than the speed problem. Light has all the characteristics of a wave, which is a proven fact. But light also moves through empty space (vacuum) where there is nothing "to wave" in. According to physics and logic, only particles (or energy quanta) can move through a vacuum. Einstein solved this problem in the only way he could see. He assigned a double personality to electromagnetic radiation.

Light may have a wave characteristic if we talk refraction, diffraction or polarization, but if it has to move through a vacuum, it will be made up of particles (energy quanta), or so-called photons. Take your pick, "as is most convenient in each case". (This "convenient" part is not my joke; I found it in the *Concise Encyclopedia of Science and Technology*, Crescent Books, N.Y. Of course, I do not like this double personality for light.

Some physicists have in the past come up with the idea of an "ether", a mysterious non-physical something in space that is the carrier medium for energy waves – a something where energy could "wave in". At that time this idea made sense, because a passive carrier medium was needed, just like water for ocean waves. Then came the famous Michelson-Morley experiment in 1887, which eliminated this nice theory of an "ether". Later Einstein saw in his own way that an "ether" is not required.

So, what is my solution to all the above problems with the speed and behavior of light?

I can see only one possible way for all of this magic to work. Logical reasoning requires it. An *underlying cosmic carrier system* with the constant speed of light is the explanation that works and can be comprehended! It is an active system and has an energy content.

We have many such systems here on earth. Each one runs at a constant speed, and everything that is dropped or put on it starts moving instantly at the speed of the system. I am talking about conveyor belts, ski lifts, moving walkways, escalators, and assembly lines, just to name a few.

And again we have *carrier rays* that come from all directions, and go into all directions at the speed of light. And the width of these cosmic conveyer belts is again infinitely narrow. I postulate that these carrier rays are identical to gravity rays! They contain energy and cause gravitation as already explained, and the weakest quanta without any own power can take "a ride" for billions of years. They enter the ray at the point where the ray exits the atom that generates it. After the ride on that ray they jump off, or get stuck in the first atom they hit. But the ray continues on its way through that atom and, while passing through, leaves a minute kinetic impulse behind, causing gravitation.

This new theory must be seen in context with my other two theories about the universe and gravity. My new theory of gravity is very much connected to this one, because the gravity rays and the carrier rays are one and the same thing. My theory about the universe provides the required back-up for the other two, because only an infinite cosmos can have such a basic fabric, consisting of these rays everywhere and in every direction. And if we add the spiritual factors mentioned in the first seven chapters and at the end of this chapter, then we have our complete *theory of everything* (TOE)! Just uniting quantum mechanics with relativity will not be enough, especially after we discovered that relativity is not needed for such a picture.

Let us summarize what we have so far with this theory for the nature of light:

1. The final consequences of Einstein's Special Relativity do not apply because they are nothing but paradoxes and defy logic.
2. The gravity rays also serve as carrier rays for electromagnetic radiation. They are the basic fabric of the cosmos.
3. This system is the only logical solution for the problem of the constancy of the velocity of light.
4. At the instant that an energy quanta leaves the outer layer of an atom (is radiated), it enters a carrier ray (or bundle of rays) which exits at this point, after having passed through the source of the radiation. The "ride" starts instantly with the speed of light. Quite similar to a sack of sand dropped on a conveyer belt which starts moving with the belt instantly.
5. At the end of the ride, the carrier ray will continue through the object of matter, now working as a gravity ray, but the

"passenger" gets stuck in the first atom, or has to jump over to an outcoming ray in order to be reflected. Thousands of situations are possible, depending on the circumstances at the subatomic level.

But I have more to say about this theory. First, let us talk about the gravity/carrier rays themselves. I imagine the energy quanta of electromagnetic radiation to be oriented on the ray like a string of pearls, resulting in a wave characteristic in a vacuum and in matter. Each quanta spins or vibrates perpendicular to the axis of the ray, that is why we can polarize it, etc.

These rays must have an infinitely high frequency, which is about the same as none at all. This is for two reasons. First, it has to pass through matter without much resistance and without doing harm. Second, it must have enough capacity to carry every kind of energy frequency. (Like how many bits of radio signals an assigned radio frequency can carry.)

Because these carrier rays contain energy, they are subject to the disturbed network of rays around a body in space or what we call its gravitation. The rays are bent towards the body in space (not the photons) because the rays from the outside are stronger than the rays that passed through the body. This is much simpler than Einstein's curving space-time. It is about the same as a train in a curve. The train (rays) follows the track around the curve, not the people (photons) in the train who are taking a ride.

Where do these gravity/carrier rays originate, and how? We do not know, but maybe we can reason it out to a certain degree. First, we have to realize that every spot in the infinite cosmos is in the center. I know that sounds crazy, but it is true. Maybe the example of our earth will help. The surface of our earth is two-dimensional and infinite. You can run around in any direction, there will never be an end. Therefore, we can reason that every spot on the surface of the earth can be considered to be in the center of the surface. The same applies in a three-dimensional way to an infinite space.

If we follow this track of reasoning, then we can say that

the carrier rays originate everywhere and nowhere. This is all we can find out about the question of where. There remains the problem of how.

If we want to solve this problem, we have to take care of another problem. My theory of gravity states that rays lose some of their energy while passing through a body in space, a planet, moon, or a star. Because this happens for each ray again and again for eternity, after a certain time all rays must lose all their energy and the cosmos stands still. But this will never be! Because of the second job they have as carrier rays for the electromagnetic radiation, they will also be replenished again and again. I see it in this way: When a photon is ejected from an atom, it contains some kinetic energy. Because it "jumps" onto an exiting ray, it will transfer this impulse to the ray. (It is similar to jumping onto a standing skateboard and causing the skateboard to move.)

Therefore, the answer to the question of *how* is this: This cosmic system of rays that cause gravity and transport radiant energy is the self-regenerating basic fabric of the cosmos. It causes everything to be and keeps everything running. It is the power source for the cosmos, and an integral part of the creation of the causal cosmic intelligence.

Why does light slow down when it passes through a medium other than a vacuum? In air, water, glass, etc., light slows down very little, but it does slow down. I believe the reason is that the rays now must pass through between the atoms of the medium. Each atom, being a tiny body in space, has its own gravity field, which bends the ray a little bit. The result is a wavy trajectory that is longer than a straight line. Therefore, the ray requires a little longer. It appears to be slower, but in reality it has the same speed on a somewhat longer track that is not straight.

Between the atoms of any body of matter there is still some empty space, just as out there in the universe. When these spaces are wide enough, the rays will go between the atoms, and we call that material transparent. If, however, the atoms are too close, the rays have to pass through the atoms, the photons get stuck in the atoms and the object appears solid, not transparent.

And now the big question. Why is a black hole invisible? We know it is because even light cannot escape. Our scientists have many different ideas about how this can be. My two theories of gravity and light propagation provide a clear answer. Because a black hole absorbs all gravity rays, no rays exit on the other sides. And because gravity rays also function as carrier rays for light, photons at or above the surface of the black hole cannot take a ride when no carrier rays exit from the body of a black hole. It is that simple! This condition of a black hole was what made it clear to me that gravity rays and carrier rays must be the same.

There are a few more important issues. They are especially of great interest for scientists who are familiar with certain "proofs" of Einstein's theories.

Einstein predicted that clocks will run more slowly in a strong gravitational field than in a weaker one. An atomic clock flown at an altitude of 10 kilometers has been shown to run slightly faster than the controls at sea level, proving that he was right. How did he pre-calculate that before the jet age? I guess he interpolated between the surface gravity and zero gravity. I understand this phenomenon differently. At a higher altitude, the controlling atoms of the clock are subjected to less shielding from the earth and therefore are slightly more stabile, which in turn changes their resonance frequency. If we could have a mechanical clock of such precision, no difference would be observed. (That is what I think.)

There is the so-called "Twin Paradox of the Special Theory of Relativity". It is supposed to be a fact. It talks about one twin remaining on earth, while the other twin takes a trip aboard a spaceship at very close to the speed of light. When he returns to the earth after a two-year-trip, he learns that his twin died a long time ago of old age, while he is only two years older. His clock aboard the ship registered two years, while the clocks on earth ticked away a whole century. I believe this is plain nonsense. I have already explained why, when I talked about the Special Relativity. I can think of only one place on earth where the twin paradox is useful, and has been used many times: Hollywood!

I have my own proof that Special Relativity cannot be right. In 1997, we observed the first objects of more than 8 billion light years away, which move through space at the speed of light. According to Relativity we should not be able to see them, because their length will be zero. But because nobody out there ever heard of this theory, they show themselves in their full three-dimensional beauty – at the speed of light!

So much about the nature of gravity and light propagation. But there are still a few more words about energies and forces.

The speed of sound goes by the same rules, but here the speed is different for each medium. And sound needs a medium (no sound in a vacuum). The speed of sound is also independent of movement and speed of source and observer, but the frequency changes. The sound of an approaching airplane is of a higher pitch than normal, and after it passes overhead and flies away, the sound will be of a lower frequency. It is actually the same effect as the red shift and blue shift of the star-spectrums in astronomy.

A supersonic jet flies away from its own noise, which originates behind the nozzles of the jet. It is very quiet inside. The speed of sound is fixed for every medium, and is very different indeed in each medium. In air it is 340 meters per second; in water, 1,450; in limestone, 4,000; and in granite and aluminum 6,250 meters per second, to name a few. And sound contains energy.

Electricity is another mysterious energy form. We know very much about it, but we do not know *what* it is. Electricity can be generated (actually only released) in many different ways. It is generated when a magnetic field and a conductor interact while in motion (generator), or when two materials of a different basic charge are connected by an electrolyte and exchange ions (battery), or when two opposite polarities are moved away from each other (potential electricity, lightning). There are many other ways to generate electricity, like solar cells, friction, etc. but they are less common.

Electricity is also very much engaged with magnetism. Both can trigger each other. So far we do not know what magnetism really is, but we work very well with it and we have all the mathematics for it. Maybe it is a special form of releasing the gravitational force; we do not know. But I am absolutely sure that here again the effect of attraction is an illusion. Something must be pushing!

Two other energies, both very similar, are kinetic and potential energy.

Kinetic energy is hidden in a moving mass and gets larger with the velocity. This is caused by the fact that more energy is required to accelerate this mass in the first place, like a bullet, for example. It is a very strange energy. It does not change the moving mass in anyway. We cannot see or sense it in anyway. Actually, we cannot even comprehend it. At the instant that a moving mass hits something, this strange energy is released. At very high velocities even a small mass can do much damage (for example, a bullet). A huge mass at very low speed is also powerful and, for example, keeps our planet rotating seemingly forever. I just said that we cannot comprehend this energy, but a while ago I showed how this energy can be comprehended through my new theory of the nature of gravity.

Potential energy is lurking in a mass that is at rest, but under the influence of a gravitational force. At the moment this mass is released into a free fall, this potential energy becomes kinetic energy. For example, a tamper or the brick that slips out of your hand and lands on your toe.

All of the energies can be converted into the energy that seems to be the most ubiquitous and important: heat energy! It is the energy of moving (oscillating) molecules or atoms. Heat energy likes to travel. It always travels from an area of higher temperature to a colder area until balance is reached. It does so by conduction, convection or radiation. No heat at all is identical with the lowest possible temperature, minus 273.16° Celsius (centigrade for some), or absolute zero degrees Kelvin.

There is a hypothesis in the world of astrophysics about the "heat death of the universe". This would mean that God was a bad planner and designer. No way! Such a thing will never happen; the whole cosmos is designed in such a way that heat will be generated again and again. It is a never-ending cycle of converting, transforming and regenerating of all forms of energy.

One last word about the radiating energies. All of them are very strange indeed, and actually cannot be comprehended. For example, let us look at heat radiation. The sun radiates heat towards the earth. There is heat on the sun, and heat is released when the radiation arrives on earth. But there is absolutely nothing to see or to sense in anyway in the vacuum of space between sun and earth. It is cold out there, very cold! Heat, defined as oscillation of molecules, cannot exist in a vacuum. Where is it? Only if the radiation hits a target of matter (the earth, a satellite, or a comet), then the heat is released. And all the other radiating energies are odd in the same way. The cosmos is weird! (Of course, my theories of gravity and light show where it is in a vacuum. It takes a ride on the gravity/carrier rays until it hits an atom!)

THE ASTRAL WORLD

The astral world is the "other dimension". It is where we go to, or better, go back to, when we "die" here on earth. Within the astral world is, for most of us, our permanent home for as long as we are required to do our karmic reincarnation cycles on this planet. (A few go straight through to the higher ethereal realm because they are finished with the earth and do not have to reincarnate, or they were only incarnated as a master, etc. in order to help us.)

This astral world is not of a purely spiritual nature. When

we die here we incarnate there into an astral body. This astral body was and is entirely ours; only the mental attributes of our soul (consciousness, memory, etc.) make the transition from the physical to the astral world.

The astral world is also a kind of physical plane, similar to ours, and exists parallel to our world. Astral objects can be identical counterparts to things in our world. In the physical world, it is impossible for anything physical to occupy the same space that is already taken by another form of matter. But astral forms can occupy the same space that is already taken by a form of physical matter. This means that when we make the transition from here to there, we do not have to change location very much; we just go from one level of existence into another (or as many prefer to say, from one dimension into another). This may go to extremes. For example, when a man who is mentally very much attached to a house that he is building, dies before he can finish it, he may find his house in the lower astral world and continue to work on it. This is not a joke; it is not funny! More than one reliable spiritual channels (most are not reliable) transmitted such conditions (including Edgar Cayce).

The structure of the astral world is something between the physical world of matter, and the ethereal realm of pure spirit (also etheric). Light and color are the main components, and everything is of a finer matter than our world. The vibrational level is so high that we cannot even conceive it. It is impossible to detect anything astral from our level. This is bad news for the many materialistic-oriented scientists who accept only proof under laboratory conditions.

Souls who are on their way to the astral realm, but mentally still very much earthbound, can remain in a kind of in-between state at the lower of the three levels of the astral realms. In fact, they may not be aware of the fact that they have died. These entities can sense everything in our physical world and sometimes even interact with us a little because their mind is desperate to make contact and be recognized (so-called ghosts). This can go on for a long time by our standards of time because the astral realm does not follow our concept of time. Everything is *now* but in sequence. For us it is almost impossible to conceive or understand the astral world.

The existence of the astral world is required by the law of duality. There are three levels. The lower level is for the earthbound entities as well as entities who require "shock treatment" under hellish conditions. For example, people who commit suicide. It also contains people who hate or think of revenge, or those whose ownership of material things carried over from the past life on earth. But there are also some entities who remain in this lowest level on purpose, by their own free will, because they want to help people who are still here. Most of the time these entities do not stay long in this level. When the "job" is done, they move on to the higher levels.

The second level is the main level. It is for all the souls who are not suitable for the upper astral. The majority of souls remain in that level, which is in reality our home. (In a way we are like actors, who play in different operas on various stages. The stages are the incarnations, and after each play we go home – to the second astral level.) In this level, we receive consultation and advice, meet old friends, do some work and learning, and finally plan and arrange the next incarnation. This next incarnation is almost entirely dictated by our karma, the spiritual law of cause and effect. But it is also influenced to a high degree by what we learned and accomplished in the past life as well as by desires of the last life. We cannot take anything with us, so goes the saying, but we can! Everything we learned our knowledge, goes with us and helps to shape the program for our next life on earth. This next life will begin very much at the same level the last one finished.

The third and highest level of the astral realm is for the souls who live there permanently. These are the souls who have no earthly karma. In fact it is the level where our teachers have their home. Their job is to teach the souls who are in the second level of the astral, and to watch over incarnated souls on earth as guardians. This third level is free of any evil, hate, or ignorance; it is just full of love and bliss.

Everything we see in the second level of the astral world is exactly as our mind expects it or wants it to be when we arrive after the transition. Nothing is in a certain way; we see it our way! A good example is the entity who takes care of us right after arrival. For devout Christians it is Jesus most of the time, for others an angel. For a Buddhist it is Buddha or a beloved teacher. For a Hindu it is Krishna, Brahma, Shiva, or any other of their many gods. A Moslem may imagine the presence of Allah. And, of course, for the entity who leaves the earth with a very bad conscience (murder, etc.) it will be the devil or some other evil being. (But there is no hell.) And the atheist will wonder what is going on.

This means that we should never take literally what near-death experiences (NDEs) of others tell us about how it looks over there. It looked that way for them, but for us it will look in a way that fits our mindset. But this condition of "as we think" exists only right after arrival (to make it easier); then slowly the reality sets in. So, what is real?

It is a long story, so I will give only the "highlights". For more detailed information read the many books that deal with the topic of "Life between Lives".

After the being of light and our guardians have helped us to understand what is going on, we will have our life review. Everything we thought and did is shown to us, and we even feel how others felt by what we did to them. We feel their joy or their pain! After that we are asked if we are satisfied with our past life. Was it a gain or a loss? (Spiritual!) Did we follow the planning we had done before we were born on earth? Did we get lost or make a mess of our last life?

Then another being leads us into a cleansing area, a kind of spiritual shower. There we will be cleaned of all the emotional "dirt" from our last life on earth, so we quit worrying about the things on earth. We do not care anymore. After that we suddenly recognize this being as an old friend, as the entity who served as our guardian during our time on earth. He (or she) then leads us into a group of people who are our spiritual "clan" (between ten and fifty friends). At this moment we are *home*! We know all of these friends by name, and have been together for millions of earth years. In fact, we had incarnations together with all of them again and again. We will miss some and ask their whereabouts, and we will be told that they are incarnated right now.

How do we look? We are beings of pure energy, looking like a point of light, but each with its own personality. Within such a group we are all at the same level of development, and therefore have all the same color of light. It starts with white for the least developed, then orange, yellow, and finally blue, with the deepest violet being the highest level. Each group has its own teacher from a higher level.

Here I would like to interrupt with a question. While reading this, did you have a feeling of being homesick, of knowing that this is the truth?

What do we do while with our group in the second level? We learn! We learn everything we want to learn. We decide what we want to learn; there is no fixed program. We learn by our own free will, because we have to develop on our own; nobody can do it for us. There is a lot of teasing and laughing, and we have a lot of fun. We do not want to incarnate again. It is much better there, at home!

But after a certain time, from a few weeks to hundreds of years (earth time!), we see that we have some karma left and that we must work that out. Our mind is getting ready for another round. We make our plans with other souls from our clan who want to reincarnate with us. The whole next life for each one will be planned from birth to death. But the only absolutely fixed decision will be the parents we select. Everything else in our life will depend entirely on our free will. Only opportunities and coincidences will be scheduled. What really happens is up to us – it is a question of how well we listen to our subconscious memory, where the plan is stored. Will we follow the feeling of our "gut", or will we reason out every decision?

Every soul prefers another time for the final connection to the fetus. Of course, it has control over the development from the time of conception, but the final occupation (incarnation) is different. For some it is a few months before birth, for others just before, and many right after birth, because they do not want to experience the trauma of the birth procedure again.

Here I would like to repeat something I mentioned earlier in this book because it is so important. A baby *always* knows and understands much more than the parents think! Very often it knows even more than the parents. This is knowledge of the soul, not of the brain. The brain is only prewired and has to be programmed through learning. While the brain learns and is under development, the soul-knowledge slowly fades away until complete amnesia of the spiritual knowledge occurs, normally within the first three to five years. A baby that is only a few months old can still read our minds (like in the astral) and is often very amused about the beliefs of the people around it.

All three levels of the astral world, especially the upper two, have one thing in common. Fights, crime or wars do not exist in the astral realms, and are impossible because of the structure and closeness to the ethereal realm.

THE ETHEREAL REALM

Now we are almost at the top. This is the realm of the souls and free spirits, including their consciousness, free will, desires and emotions. It is a realm of pure thought energy, also called the etheric. The inhabitants are the ethereal beings.

The souls are the main group of the etheric. All have full control of all lower levels, all the way down to the manifestation of material matter, physical as well as astral. They perform works on a scale we cannot conceive; their power is almost absolute. Some are only responsible for a certain part of a planet and the life on it, others for a whole planet or a solar system. And so the ladder of responsibility goes up and up. There are leaders of whole star systems, galaxies, galaxy clusters, and finally the universe. Of course, it is never one entity alone that is in charge; each federation is under the leadership of a team made up of the members. It is actually a hierarchy of the cosmos. (As in heaven, so on earth! Sounds familiar?)

*They are the souls who operate and govern the universe*! The universe is theirs! That is the primary topic of this book, *The Universe and its Souls*!

All beings of the etheric are interconnected within the causal cosmic intelligence, God. Each task they perform is in reality an act of God, because they are part of Him. (Like everything else.)

*Him*? Not her? Like God, all beings of the etheric are not of one sex; they are both within one entity. The male state is an active one. Male is doing, building and destroying. The female state is a passive one. Female is loving, sustaining and nurturing. They are just two polarities, the two sides of one perfect being. The fact that both polarities are united in one perfect being is very nicely shown in the famous Tao symbol of yin and yang. I guess that many people who wear this symbol as jewelry do not know what this symbol really stands for.

As stated earlier, sometimes these pure souls incarnate on earth in order to help us find our way back (the masters). They always incarnate as a man, because the role they have to play is an active one.

Each time a soul wants, or has to incarnate into a physical body, there is the problem of the two sexes the biological bodies require. The solution is simple. Every soul must separate into a male and a female personality, and then occupy two bodies as long as their cycle of incarnations will last. We call them twin souls. (This is the actual reason why earth's population is always about half male and half female, even after an "overkill" of one sex in a war.)

Towards the end of a long series of incarnations, both parts learn about this fact and begin to search for each other. After they find each other, they continue to incarnate – always together – until both no longer have to incarnate. Then they are one again with God, in the body of an ethereal being. (The best-known example is the mother of Jesus; Mary; who is the twin soul of Jesus. Both are equal in development and have the same powers.)

Because the etheric is nothing but pure spirit, being part of the causal intelligence, the beings of this level are in fact individualized cosmic mind energy. We cannot comprehend the way of how they perform their tasks. We know what they do; but we do not know how they do it. But we can reason.

Generally it goes like this: First, they see and accept what has to be done. Next, they develop an idea of how to do it. Then they will have the desire to do it, know and give the command to the causal mind to execute the task. Space and time being already provided since eternity, all the causal mind will do is set up the required manifestation, change, or elimination into action. The causal cosmic mind is actually the servant of the ethereal beings, which are in fact cells in His body.

Because almost nothing can or should be fulfilled instantly; especially if the subject is a planet or a sun, the requesting ethereal entity also has to provide patience – and in large quantities indeed for the evolutionary processes which will require millions or even billions of (earth) years. Here we have an interesting parallel. In one of the Cayce readings the question was, "What are the requirements for the existence of the physical world?" The really surprising answer was, "Space; time, and patience." Space and time were no problem, but patience? It appeared very odd at that time. I hope that this paragraph will be of some help.

From all of this we can learn one important lesson. We must understand that our souls are in reality citizens of the etheric. When we pray, we must not beg for something. We should request it, even to command in certain cases. Why? Because in fact we talk to our own higher self in the ethereal body, or the superconscious mind. And of course, we have to have the patience to have it done properly and at the right time. But do not expect any result if the request is selfish or not for a good cause. It is useless to pray for help to win a war; war is always a bad deed! There is no such thing as a good or a holy war! (But millions do pray for victory and then are disappointed.)

If the request of our prayer is good and positive for our karma, or good for another person we pray for, it will be fulfilled. If it is not good for us or the other person we pray for or if the prayer is for an immoral purpose, it will be discarded. That is why prayers are often not "heard".

We may be able to understand that ethereal entities have only to think of something, want it done, and it will be done. But how the causal mind of the cosmos does it, we will never be able to conceive of here on earth.

Not all souls are busy as caretakers of the cosmos. Some are assigned to take care of others, or do so of their own free will.

They watch over entities who are incarnated in the physical world and have made a certain progress already (the effort must be worthwhile). These incarnated entities are followed by these guardians through all incarnations and get advice from them when discarnated between lives in the physical world. (Guardian angels for most people.) They watch and make sure that nothing will happen that could interrupt the final development of the incarnated soul. (Mine did a superior job so far and even helped with this book.)

Another group of the etheric are the so-called free spirits. They are in control of whole groups of animals or plants. For example, they are the ones who tell the grunions how high the tide will be on a certain day a few weeks from now, so they will bury their eggs at the right distance from the shoreline They are also in control of all the migrations of fish, birds, insects, and land animals. The total extent of the work of the free spirits is beyond imagination. It is interesting that some people who can see other people's auras can also see some of the free spirits, especially the spirits in charge of plant species. They call them nature spirits. (The Fairy stories originate from this. Many small children can see them and even play with them. The boy Edgar Cayce played with many of them behind his house and in the woods.)

So much about the beings of the etheric. But another very important part of the etheric are the *Akashic Records*. Edgar Cayce mentioned the *Akashic Records* as the source of his life readings. I think it is very interesting that Edgar Cayce did not know anything about Hinduism and nothing about Sanskrit. But in his readings he said Akashic, and Akash is in Sanskrit the word for "the ether, on which God projects His creations". This small but important fact is just another very convincing verification of Edgar Cayce's readings.

These *Akashic Records* are a very strange thing. We know what they are and what they do, but how they work and what the real purpose of them is, we can only speculate. I believe that we cannot comprehend them at all. They contain every thought and every act of every life form throughout the history of the cosmos, which is eternity. In addition, these records contain everything that ever happened within the cosmos. It seems to be divided into an infinite number of sub-records, or volumes, or "books". One for the earth – that is where we can tap in – others for the solar system, and so on. The records of the whole galaxy, for example, would be useless for us, because we do not know what they are about anyway. The *Akashic Records* are not in writing, or on tape, or on a disc; they seem to be a kind of thought recording of the universal mind.

After we die here on earth and "crash-land" on the other side, we will first have a change-of-dimension shock. But quickly we will recover and feel normal again, ready for new impressions. Then we will be shown the last section of the long "movie" of our existence, our part of the *Akashic Records*.

This section will replay the life we just lived, including every thought and action. But worst of all – for some people – we will experience not only the joy we brought to others, but we will also feel the pain we inflicted on others. And we feel it just the way they did! But there is an exception, and indeed a very important one. We will not feel the pain we caused others, if these other people have forgiven us! And the karma it caused will also be waived. This is a parallel law to the law of karma; it is the law of grace. Jesus repeatedly spoke of this fact and lived as an example of it.

From time to time, some people, like Edgar Cayce and many others, have the ability to move their consciousness into the etheric and tap into the *Akashic Records* while here on earth. From these sources we can learn many more facts, and more importantly gain knowledge, than from any scientific laboratory.

THE CAUSAL COSMIC INTELLIGENCE, GOD

This realm which is the highest and contains everything there is, is called by many names: God, Causal Force, Universal Mind, Cosmic Intelligence, and many others, depending on language and religion. We know that none of them is right, because the Vedas say, "If you cannot comprehend Him, you cannot name Him!" But for the purpose of just talking about Him, we have to invent and use a name. "Causal Intelligence" seems to be the best name. "Causal" because it is the cause of everything there is, and "Intelligence" because it must be an intelligence according to the reasoning of Immanuel Kant (and others). Let us talk about this man.

Generally, it is agreed that Immanuel Kant was the greatest philosopher who ever lived. He solved many problems of physics, metaphysics, astronomy, and moral laws, using only his amazing power of logic and reasoning. He lived from 1724 to 1804 in the city of Königsberg, Germany (today Kaliningrad, Russia). He never left this city in his whole life. I will quote from Kant's famous book *Critic of Practical Reason*, which he published in 1788. It is truly amazing what he did just by reasoning. After many pages of "pre-reasoning" he arrived at this conclusion about the existence of a causal being or force. I quote:

> The objects of the material world are fundamentally unknowable; from the point of view of reason, they serve merely as the raw material from which sensations are formed. Objects, of themselves have no existence, and space and time exist only as part of the mind, as intuitions by which perceptions are measured and judged… Therefore, the *summum bonum* is possible in the world only on the supposition of a supreme being having a causality corresponding to moral character. Now, a being that is capable of acting on the conception of laws is an intelligence (a rational being) and the causality of such a being according to this conception of laws is his Will; therefore the supreme cause of nature, which must be presupposed as a condition of the *summum bonum* is a being which is the cause of nature by intelligence and Will, consequently its author, that is God!

Immanuel Kant also developed a theory of how the solar system was formed. It is still the best and most accepted theory we have today. And he also believed in reincarnation; it was the only answer that was logical to him and made sense. Anyway, I think that this quote is a good introduction for a discussion on the causal intelligence.

Why did God do what He did and create a cosmos? Because of the seriousness of the subject, I have to employ the knowledge of a group of people who knew much more than all of us combined. The problem is that nobody knows who these people were. They wrote the Vedas many millennia ago. As far as I know, the Vedas are the only source for an answer to the question of why God did what he did. (Edgar Cayce said that God created us for companionship, but that is not satisfactory, not enough.)

I cannot quote the Vedas because of space limitations. The Vedas are immense! But I will try to give an extract of the meaning of the part that is of interest to us, in a language that can be comprehended.

*Before the beginning* God was in a perfect state of absolute balance, peace and love. He was eternal and limitless, His powers were infinite, and His ideas and desires were at rest. All He did was being perfect, nothing else.

Being perfect is nice, but it is not only boring, but also useless; because no action, no motion, no manifestation, and no learning occurs. In short, progress was not possible.

He then had the desire to change that. First He created out of Himself individualized spirits, souls, for companionship. Each of these souls, an infinite number of them, received a free will just like His, and was pre-programmed to serve as co-creators and eventually to come back to be one with Him again. And naturally, each soul inherited His desires for companionship, to design and create, and also all of His powers. Each soul was an identical copy of Him (in His image), so identical that each one could believe that he had been first. His own consciousness came out of balance, out of the state of rest because of all the others. The result was action!

Then He developed the plan of a cosmos so the souls had a purpose and something to do. He provided space, time, matter, and all energies and forces. In order to be permanently informed about everything old and new in the cosmos, He set the *Akashic Records* system between Him and all other things. Everything the souls did, and everything that happened in the cosmos, went instantly into these records. He connected Himself to the other side of the *Akashic Records* in order to be always informed about everything there is or happens in His cosmos. At all time He is in full control and He progresses by learning all that is going on. The rest is eternal history.

That is all I can say about the causal intelligence, God. There is no way to comprehend Him. That is why we know so little about the source of our existence.

So, what are these six groups we just talked about? Science? Religion? Mysticism? Fiction? Let us look at them from a different angle. Swami Sri Yukteswar, the guru of Paramahansa Yogananda, wrote a book in 1894 that had the title *Holy Science*. He wrote the book following a request of his guru, Babaji. Even though these masters grew up in Hindu families, they became freethinking spiritual masters. Yukteswar was the "incarnation of wisdom". I think that the title of his book would suit these six levels of existence very well. In fact, even for the whole content of this book, because *all of this is science*!

His book actually had the same purpose as this one: To show that most of our religions teach basically the same tenets; that there is *one* God, that the rules are almost the same, and not just a belief or faith, but knowledge has to be the goal. He also made clear that science and religion are so strongly interwoven that they cannot be separated.

Now, one hundred years later, it is the purpose of my book to make clear that Einstein's relativity, Planck's and Heisenberg's quantum mechanics, and Newton's laws cannot be united if each one goes alone, and if the existence of a superior causal intelligence is rejected by ignorant prejudice such as "that is not scientific".

At the very moment a scientist has the courage and accepts the fact that the thought energy of the causal intelligence is the basic force of all science faculties, he will make great strides ahead! Astrophysicists in particular, must understand that our universe, with its billions of galaxies, each with billions of stars, does not exist for our entertainment only. And some even ask the question, "Are we alone?" Such a question defies any logic. Logic tells us that just by its immense size, and the trillions of stars, the universe must be teeming with life. It has its organized and regulated hierarchy of leaders and masters. Everything has a purpose and nothing exists for our entertainment only. The universe is exactly like nature here on earth: nothing is without a purpose. Even an exploding star, a supernova, has a very important purpose. It is how the universe creates the required heavy elements. Our planet and our bodies would not exist without them.

Some scientists already think about these issues and are ready to accept these facts. But they are only a few and keep it to themselves so as not to be ridiculed by others. I wonder why Einstein seems to be the last one who openly admitted to being "cosmic religious". During the last century, science has become increasingly materialistic. Why? I think it is because during this century science had to serve the military needs of the nations more and more, until it almost completely depended on grants from that financial sector. It is time to return to the moral levels of the nineteenth century.

It is my belief that Einstein's idea of being "cosmic religious", to look in awe at nature and the cosmos and to admire its creator, is true religion and separates itself automatically from the man-made, money-collecting churches of institutionalized religions. We are all part of the great causal intelligence and we do not need an agent between God and us. We can communicate directly, because He is in us, and we are in Him. Just believing what the church says, or having faith, will never be enough. *Knowledge* has to be gained. Dogmas and ceremonies will never do it for us!

But many Edgar Cayce readings said that if you like your church and it serves you well, and makes you feel good, then do not give it up because of this new knowledge of the truth. Only if you never cared much about, or believed in a faith, you can stay away, still knowing that there is a God.

COMPLETION

Before we close this chapter, I have a few more thoughts about the universe. This last chapter has provided some fundamental facts which make it possible to answer a few questions that are still unsolved problems in science. (I could not touch these topics earlier.) Some are real big problems, because they involve strange physical paradoxes.

Did Edgar Cayce say anything about UFOs? In a way he did. In reading #1616–1 he said, "The entity was among… the Mayan experience… when there were those that were visiting from other worlds or planets." This tells us that they did so in the past, but not anymore, because they know that now we have the habit of shooting at everything we do not know. They

# A THEORY OF EVERYTHING OR "QUANTUM-RELATIVITY"

cannot help us anymore because of our ignorance.

Why do the outer ridges of our galaxy orbit much faster than they should? The only idea our scientists have now is to search for an invisible mass that could account for this paradox. But so far it has not been found. My theory of the universe does not need that extra mass.

Each one of the over 100 billion solar systems (now some say 400 billion) of our galaxy operate similar to an atom, the planets being the electrons. Just as atoms do not orbit around each other in a piece of material, the solar systems in the chunk of matter which is our galaxy, do not orbit either. They are interlocked in a similar way as atoms interlock. The gravity and magnetic fields throughout the galaxy make them behave like a cluster of magnetic steel balls. They cling together. The result is that there are no orbits of stars. The whole galaxy rotates as one huge wheel with some slipping towards the rim. Therefore, just as in a wheel, the velocities at the rim are the highest. (Orbiting is impossible anyway if we consider that at the beginning of a galaxy there is no dense center to provide for orbiting.)

Exactly the same applies to the universe; there is no orbiting either, but rather rotating like a big wheel. That is why the velocity at the rim finally reaches the speed of light.

Why are the distances between the galaxies so unbelievably large? While the spaces between stars within the galaxy already stretch our imagination, the intergalactic spaces, the huge voids between galaxies, are immense, almost impossible to comprehend. The reason is identical for solar systems within a galaxy, and for the galaxies of the universe. Each galaxy began as a huge cloud of matter (dust) in space that was larger than the final product, the galaxy, by magnitudes. Billions of "lumps" formed within these clouds while the huge cloud rotated and began to shrink towards the center. When the cloud had condensed to about five to ten times the size of the final galaxy, the first lumps condensed into stars. Finally, the billions of lumps became stars and complete solar systems appeared. Because the galaxy started out as a cloud hundreds of times the diameter of the final galaxy, it left behind this immense void in space. (The globular clusters that surround each new galaxy are nothing else but stragglers that still have to join the club.)

Why does everything in the universe rotate or orbit? How did it start? It is the so-called whirlpool effect. If a sink is filled with water and absolutely still, it will begin to rotate counterclockwise when we carefully open the drain in the center. South of the equator it will rotate clockwise. The reason is a minute difference of the surface velocity of the earth. Because of a microminiature difference in the radius of the earth, the southern end of the sink moves faster in an easterly direction than the northern end. The water is so unbelievably sensitive to this micro difference that a counterclockwise rotation occurs.

The situation in space is similar, but everything is even more sensitive than our water, because it is in a vacuum. The whole universe is one big vortex of energy and forces – and it rotates. Each object in the universe, a galaxy for example, is subject to a higher velocity towards the outside, and the whirlpool effect sets in. And as a galaxy condenses towards the center, it rotates faster (like an ice skater who does a pirouette). Stars and their planets are subject to the same kind of energy vortex within the galaxy and begin to rotate.

Some of our astronomers are very busy mapping the way galaxies are distributed throughout the universe. (It became important when Margaret Geller was the first to see the so-called stickman, a strange group of galaxies that looks like a stickman.) So far the result is one huge question mark. Why does the universe appear so profoundly different from a universe as it should be *a la* Big Bang? Back in Chapter IX in my new theory of the universe, we can find the answer in the way matter in the middle of the universe is replenished.

Because nothing is evenly distributed and probabilities are the rule, the universe looks chaotic! We see single galaxies, galaxy clusters, superclusters, galaxy "walls", streams of galaxies and galaxy systems. We also observe lots of gas and dust everywhere. Between these conglomerations are immense voids of different sizes. According to my theory, the universe has to look like that; it cannot be evenly distributed. (Forget that nice picture of the microwave background.) The only way I expect it to be "organized" is in the form of concentric ring sections, made up from billions of galaxies; and caused by the rotation of the universe. A spiral structure cannot be expected because the universe does not shrink, it is expanding.

What about the quasars? What they are is explained in Chapter IX, and how they work is explained in this chapter. But I see possibilities and probabilities in connection with quasars that may explain some of today's problems.

When a quasar "is over", there will always be a remnant from the larger one of the two black holes who had that rendezvous in the outskirts of our universe. This leftover could be a small black hole, a cloud of large and small boulders, or just a cloud of dust or even gas, depending on the size and the "procedure" of the quasar. Now, if this remnant is of antimatter and keeps moving towards the center of the universe, we can expect hundreds of different effects if it hits one of the matter objects of our universe. For example, an antimatter cloud of dust could be subject to the gravity of a galaxy core. There it has to pass through the cloud of electrons that surrounds every galaxy. Annihilation occurs and gamma rays of a specific wavelength can be observed. To us it looks like a stream of antimatter jetting out from that galaxy; that is an illusion. The possibilities of unexplainable observations are endless.

One last word about this chapter. Many facts, observations, and accepted theories are mixed with my own theories and intuitions. I am sure that some points will eventually prove to be off the target, especially where measured values are involved. But with my opinion about the superior intelligence I am not alone. Albert Einstein said, "Everyone who is seriously involved in the pursuit of science becomes convinced that a spirit is manifest in the laws of the universe!" I think that too many scientists are *not* seriously involved in science with their mind because they reject this direction of thought by Einstein.

I hope that this book will at least be a useful base to build a more detailed picture of the universe and our purpose in it. I have done my best to eliminate many of the exotic ideas in science that disregard reality (like wormholes, time-machine-style black holes, cosmic strings, space loops, hyperspace, ten geometric dimensions, small baby universes; and many other strange ideas, some from scientists with great names).

The cosmos is not only matter, energies, forces and spirit, it is also based on logic and common sense!

That is the way God likes to do things, and so should we!

# CONCLUSION

> Discovery consists of seeing what everybody has seen and thinking what nobody has thought.
>
> Albert Szent-Györgyi

The chapters in this book have covered many questions, from religion all the way to physics and cosmology. Actually all of them are topics of man's eternal quest: Why does the universe exist? How does it work? Why do we exist? What is our relationship to the universe? What is the purpose of the earth, and why are we here? Will we survive death? If we do, what does it look like after death? Will we be born again? Is the whole system fair to us or cruel? What are gravity, electricity, or light? It is a long list. And as always in science, each answer results in many new questions. The following is a condensed overlook of what we have covered. Each topic in this book is required for a complete picture of the universe and ourselves.

Many of humankind's problems could be solved simply through common sense and logic. Logical reasoning is a tool for research! Sometimes, when you get stuck, an entirely new viewpoint may be required to arrive at a solution. More than once I expressed my strong belief that the only way to get out of a dead end street is a U-turn. And many faculties of science *are* stuck in a dead end! They are idling along in a patchwork of exotic ideas.

We still have a long list of things that we will never know the answer to. Our brain is not designed to fathom the details of their structure. The best known are space, time, gravity, light, electricity, kinetic energy, heat energy, magnetism, and best of all, infinity. The strange fact is that we have huge libraries with books full of natural laws and mathematical equations for all of these strange things. We know very well how to work with them; our technology is proof of this. As a matter of fact we have collected so much knowledge about all of them that nobody can know all of it – but we still do not know *what* they are. It could be embarrassing if we allow it so to be.

Then I added another list of things that are proven facts, but still are not accepted by science. And again we do not know what they *really* are. We talked about out-of-body experiences, thought transfer between people and animals, prophecy, clairvoyance, spirit, mind, consciousness, and many others. All are facts of a non-physical nature, things we cannot sense, measure, test, or prove in any so-called scientific way.

To top it all, we know for almost a century that matter, as we see and feel it, is only an illusion. All matter is made up of atoms. Atoms are made up of even smaller units of vibrating, rotating, or orbiting energy. And there is much empty space within an atom and even more between them. Matter is a special form of energy! This is not easy to comprehend, if at all.

The cosmos is not a peaceful place with only warm suns and nice planets. No, it is very violent in many places. There are collisions of whole worlds – explosions, supernovae, expanding and collapsing suns, black holes that devour everything, and the quasars, which are the ultimate infernos. But all of this has a purpose in God's plan, and the souls are in control of everything that happens in the universe.

On the other hand, the cosmos is full of peace and unconditional love. The things that are not of matter but of the spirit, are much more important than anything physical, including our bodies. Every soul matters and cannot be destroyed, but it does not matter much if a whole galaxy is eliminated. Immanuel Kant saw it right when he stated that from the point of view of reason, the objects of the material world serve merely as a raw material from which sensations are formed – or in today's language we could say that the whole physical cosmos is no more than just a playground for the souls. We must realize that the spiritual realm is the reality, while the physical cosmos is only an illusory manifestation of energy into matter.

If we think about the contents of the chapters in this book, we must admit that we live in a very strange and weird world. It also becomes clear that science as it acts now can never solve the ultimate problems. And religion cannot do it either. Both have to merge into a new era of total science!

Scientists must become open-minded about every aspect of life. They must forget their entirely materialistic orientation. A good start would be to accept and respect other faculties of science. Doing so, they will begin to work together instead of against each other. Then they have to widen the horizon of their general knowledge, get out of the trap of super-specialization, and accept spirit as the *foundation of science*, because that is the cause of everything. Of course, this means to accept the existence of a causal cosmic intelligence, God!

Religion must also change. Most institutionalized religions are of little use for the soul's development (which is the purpose of an incarnation). Some are so dogmatic that they even hinder the development of the human souls into a higher spiritual level. Almost every religion and their sects claim to be the only right one, and their members believe so. The word tolerance seems to be unknown. Of course, such thinking is nonsense and against all logic. Some have a priesthood that is so dictating and dogmatic, that it stops the spiritual development of the members in the tracks. I am talking about religious regulations that are in contrast to normal human feelings, emotions and independence. Ceremonial rituals, pomp and money do not help anyone's spiritual progress. Congregational buildings make only sense if used as a meeting place where knowledge and ideas are exchanged. Is there anything left? What should religion be?

The word *religion* has its roots in Latin and means to be a "way back". Back to what? It means back to the creator, to the real self. True religion is the knowledge of who and what we are, and what the purpose of our existence is. It is of little help to believe in something unknown, have only faith, or what is the worst of all, to fear God. (That means to fear ourselves because we are part of Him.) We must know that we have a free will, to be applied in the right way. Religion is to *know* that this causal force is behind all things, in all things, and *is* all things.

Then we understand that this causal intelligence set irrevocable laws for the cosmos. Physical laws for the world of matter, and moral laws for the souls, based on love. If we live by these laws, all will be fine and dandy. Just one hour a week of so-called worship does not do any good. God does not want to be worshipped in such a useless way. He wants us to live by His laws all the time, every minute. That is the only true worship: *Doing* what we souls are supposed to do!

We do not have to call this causal intelligence by any name. There is no name. We have to know that we are part of this cosmic force, and live by the laws. That is all! That is what religion has to be, to be "cosmic religious" as Einstein said, or the way Edgar Cayce formulated it: "Be still and know." If you prefer it a bit more sophisticated, go by the words of Paramahansa Yogananda: "If you know that you know, then you know all that is to know."

Many people believe that this life is a one-shot deal. They think it is all just chemistry. They do not see the signs that clearly point to a great omnipresent cause of it all. Therefore, they see no reason to bother about moral issues. Many of these people follow at least their conscience, but too many do not and provide hell on earth. (Heaven, hell, and a devil are the invention of Egyptian priests thousands of years ago. Subsequently, many religions adopted the idea, because it worked well by scaring the people with hell and promising salvation.)

There is a movement at this time, which has the focus on an open mind and spiritual orientation without an organized religion. The many books that deal with this strong movement are in the "New Age" section of our bookstores. The people of this movement understand that the cosmos is ruled by two contradicting forces; love and ignorance. Both are very powerful.

But there is a negative aspect of this movement. A great number of people misunderstand this movement and go into silly extremes. They are a new religious group that deserves only one name: Freaks! They are the people who make the New Age look foolish by performing silly rituals, witchcraft, satanism, or claiming to be psychic on a number 900 phone service. I always wonder how many people believe in this nonsense.

Thousands of years ago, the people had to be advised in detail for everything. So they got the Ten Commandments, or all the other basic rules of the various religions. Then, about two thousand years ago, the people had better knowledge and were ready for simplification. So Jesus introduced the new version: Love your neighbor and do unto others, as you want them to do unto you! It is very simple. Whoever lives by this advice automatically fulfills the Ten Commandments.

And now, two thousand years later and with all our new knowledge, we have the ultimate simplification of advise from many sources, because now we can understand the full truth. This new simple law is: KNOW YOURSELF!

This is all! And this is what this whole book is about: To know that we are individualized spirit and part of God. To know that we are on earth because we must find our way back to Him. To know that besides the physical laws for the cosmos, there are moral laws of cause and effect for the souls, karma. To know that because of our karma, we have to reincarnate again and again until we rebuilt our soul's character back to ethereal standards. To know that there is no other way!

If we know all of the above, then we truly know ourselves and understand that we have to live accordingly. If not, we will need a long time to make it back. Edgar Cayce's "Be still and know!" means the same. In reading 440–11 he gave this excellent advice, which in fact applies to all of us: "The knowledge of self... is only valuable or constructive when applied in the experience of self. Hence, get busy!"

After this short review of the philosophy of this book, we approach now the physical cosmos. We have found out that the souls are responsible for the care and maintenance of the world of matter. It is their job. Even when incarnated, we have a limited responsibility for our planet. Because our souls will someday rejoin the ethereal "workforce", it will be good for us to understand the universe as much as possible here on earth.

Because I found out that some of today's theories in physics and cosmology are wrong, I introduced my new theories about the nature of the universe, gravity, and the constancy of the speed of light. These new theories together with the philosophy we just had, are my picture of a Theory of Everything, or TOE. It has to be a conceivable picture and not only mathematics!

I have also tried my best not to add any other item to what I call "extreme exotic science", I am talking about wormholes, baby universes, evaporating black holes, and all the other abstract and bizarre ideas. My goal was to get back down to earth and develop a picture that makes sense, is logical, and can be conceived. And, of course, I have tried to solve as many of the current problems in science as possible.

This, of course, is a steep claim for just one person. I am aware of that. If Albert Einstein, Stephen Hawking, and so many other great scientists cannot solve these problems, then how can I claim to have the solution? It is, however, possible! First, all of these scientists had one crucial mistake in common. They rejected the idea of a causal intelligence God, as "not scientific". Second; my mind is that of a trained engineer; but not blocked by too much specialized knowledge which requires too much of the brain to process it. (We normally call that a closed mind.) Therefore, I have a complete panorama of the big picture in my mind. And third, all of these scientists are thinking on a scale that is too small and too complicated. And, of course, my brain is of the exact same model as Einstein's or Hawking's – human. But I use it in a different way: I try to keep an open mind, and if something is not logical and makes no sense, I eliminate it.

So, let us go forward. All the details, all the whys and hows are already explained in the previous chapters. What is still missing is a condensed summary for each of the new theories, but without detailed explanations and reasoning, which were already given.

## The Cosmos

Infinite in size and time, the cosmos is composed of an infinite number of universes, with the required space in between, which is larger than the intergalactic space within the universe by magnitudes. Fifty per cent of the cosmos consists of matter universes, the other half are antimatter universes. The linear time flow of antimatter is reverse in respect to matter, as was proven in 1949 by the great physicist Richard Feynman. This means that antimatter has also anti-time. Therefore, time in antimatter universes flows backwards from our point of view.

Because of the opposed time flow of matter and antimatter, the two cannot coexist and instantly annihilate each other when they meet. By the laws of physics, there has to be an antimatter atom for every matter atom. Antimatter cannot exist within our universe because of opposite time flow and annihilation. Therefore, alternating matter and antimatter universes are the only possible (and logical) solution. Searching for a theory and formulas of how antimatter was destroyed during the first seconds of the Big Bang is the wrong approach and contradicts physics.

From the viewpoint of the causal intelligence, the two opposite time flows cancel each other out. Therefore, the "official" time for the whole cosmos is a permanent *now*! This explains why the spiritual realms do not have time, as we know it. I think this is a very clever solution, even if we cannot comprehend it.

Of course, our matter universe is an antimatter universe for the anti-people on their anti-planets in an antimatter universe. Everything is the same there, except that electrons are positrons, protons are antiprotons, and time goes in reverse.

The fact of absolute infinity of the cosmos allows for the new theories of gravity and light propagation. All universes are of about the same size, limited by the speed of light, which is itself determined by the speed of gravity.

## The Universe

The Big Bang never occurred. The whole universe rotates like one huge wheel, just like everything else in the cosmos rotates or orbits. The universe is also a flat disc, just like galaxies are, but it does not have a bulge in the center. The galaxy cluster Virgo seems to be at or near the center; we are also close to it.

The center, and the general area around it is the main "breeding ground" for the formation of new matter, the raw material for new worlds. As galaxies form and gain mass, they are subject to the centrifugal force within the energy vortex of the rotating universe and begin a journey to the outside of the universe. Because the rotation velocity enters the formula for the centrifugal force squared, the speed of this trip to the rim is accelerating. The farther out, the faster everything goes, finally reaching the speed of light. This is the reason for Hubble's red shift, the "mother" of the Big Bang theory.

But even at these high speeds, the journey to the rim of the universe needs many billions of years. The Virgo cluster may be about 6 to 8 billion years old, our galaxy around 16 billion years. The farther out, the older the galaxies are (but not their stars). When "born", they are huge spirals with more gas and dust than stars. Later they are smaller and better organized, consisting of billions of stars and fewer nebulae. While on their trip out, they keep shrinking, first into ellipticals, and finally into small irregulars.

During this journey to the rim, a black hole at the center of each galaxy swallows star system after star system until the whole galaxy is contained in a giant black hole and is invisible (that is why the irregulars are so small). This happens while passing through the last 5 billion light years before reaching the rim. At the outer rim, approximately 20 billion light years from the center, these black holes reach a rotation speed equal to the speed of light, and the escape velocity is also the speed of light. Because they cannot accelerate any faster, they have to leave the universe and begin a journey through the vast inter-universal space (between universes) of the cosmos. In the overall average of the cosmos, only 50 per cent of all black holes escape, the others collide with incoming black holes from other universes. Most of them are antimatter. When the two meet, they annihilate each other according to physical law. This is the only way to destroy a black hole!

The product of such a collision is a quasar. Quasars are very small, only the size of a solar system, but they radiate energy in the form of gamma ray photons, and almost all other frequencies. This energy output is equal to the energy output of hundreds of galaxies combined. (I believe it is $E=mc^3$ at least.) That is what quasars are!

The quanta and particle radiation from all these quasars around the rim of the universe arrives from all sides in the center area, and therefore the number of particle collisions is the greatest there. The collisions result in the creation of the basic compounds for matter. Together with the output of supernovae and other element-creating cosmic disasters, they are the raw material for new worlds. It is an eternal cycle of creation and annihilation. No Big Bang, or baby universes *a la* Hawking that are connected to ours through wormholes. The universe is real, based on logic and clear physics, not exotic.

## Our Galaxy (and all others)

Like the universe, our galaxy rotates also as one big wheel, with some slippage in the outer spiral arms. (Elliptical galaxies rotate more uniformly.) The initial cause of all rotation within the universe is the whirlpool effect of the rotating force and energy fields of the universe.

But our galaxy does not expand like the universe, it contracts. The same centrifugal forces as in the universe are also at work trying to expand the galaxy, but there is one very big difference: The "very short" distances between the stars of the galaxy. The universe is wide open, with immense voids between galaxies; but a galaxy is dense in comparison – so dense in fact that from a distance of a few million light years we cannot separate the stars anymore. We just see a patch of light. Because of this density, gravity is stronger than the outward forces and the galaxy slowly shrinks.

Our "home galaxy", the Milky Way, is still very close to the center of the universe, only about 30 to 50 million light years, and therefore a young galaxy. Its age is approximately 16 billion years. The globular clusters of the spherical halo around our galaxy are just stragglers who still have to join the club. They are not yet part of the wheel, they still orbit and do not rotate with the "body" of the galaxy.

While many new solar systems are still forming in the spiral arms, a black hole in the center has already started to consume the galaxy from the inside out, using its tremendous gravitation to catch star after star. Eventually, it will devour the whole galaxy and end up as one monster of a black hole.

At the outskirts of our galaxy, in the spiral arms, there is still some gravitational freedom left for some suns to orbit around a central giant star. And of course, the freedom is great enough for the orbits of planets around their stars. In ratio to the size, the spaces between solar systems are very large, even though the galaxy as a whole entity is dense.

## The Earth

Our home planet seems to be the only planet in our solar system that harbors life, as we know it, which is a science of its own, called biology. We found out that evolution is impossible for millions of reasons; but *mutations* within a species are a fact. We also saw that creation as described in the scriptures of the various religions do not make logical sense either. But we found the solution: The *design* of species and then seeding of the earth is the answer. The designs are performed by ethereal beings or people on other planets who are mentally very close to the etheric. Each young planet under development has its assigned groups of "aliens", who are responsible for designing and seeding the right life forms for every era and condition. They have to start all over after every global disaster, but always with completely developed species right away. (Evolution cannot perform such restarts every few million years.)

The solid crust of the earth is only half as thick as the shell of a chicken egg, compared in ratio to the size, and very unstable. About every 40 to 150 million years, the earth has to endure a total global disaster, depending on when the next pole shift occurs, or the next huge asteroid hits. The geological layers of the earth show that right after each one of these disasters, life bounced back with fully developed life forms. They are brought in by these "aliens"!

Our planet still has many gas chambers underneath the crust, with more gas coming up from below all the time. This high-pressure gas is responsible for exploding volcanoes, vertical earthquakes, and lifting whole mountain ranges or islands in the ocean. (It is natural gas, liquefied under the high pressure.)

Coal is the product of plants being under high pressure and temperature for millions of years, as always believed. But I do not believe that oil is the result of a similar process. It is a natural part of the earth (just as the gas), slowly creeping up from below.

Our earth is too young and not ready to harbor a super-civilization of the highest spiritual and scientific level because it is not stable enough. But it is not only good enough, but perfect, to serve as a school for souls who wandered off the right path.

## Gravity

Gravity cannot be an attracting (pulling) force for many reasons, one of them being common sense. (How do the atoms of a body in space pull through empty space over distances of millions of light years?) Gravity has to be a pushing force!

In my new theory, "gravity rays" approach every point in the cosmos from all directions at exactly the same intensity (energy content), and they go into all directions. The speed is the speed of light. This "network" is the basic fabric off the cosmic space, the power source of the cosmos. It forces bodies in space together and holds them together. This holding force is the cause of inertia (the tendency of a mass to resist movement, and to keep moving if it moves), and it is inertia that keeps everything moving, rotating or orbiting.

These gravity rays are infinitely narrow, and every infinitely small point in space is affected by it. Because of their energy content, they are an *active system*, and not passive as the once theorized ether. A planet is of almost no resistance for these rays; they pass right through.

These rays work not only on each atom, but on the very smallest subatomic parts of an atom. While passing through, they transfer an almost infinitely small impulse of kinetic energy to each subatomic particle. If this atom, or a mass of atoms, is alone in space, the effects balance out, because it is from all sides. But nothing is alone in space. Looking at the earth and moon for example, we can see that the moon receives weaker rays from the direction of the earth. The earth is *shielding* the moon from the full power of the rays. As a result, the moon is pushed towards the earth. *Shielding* is the name of the gravity game! Because the earth shields us (weaker rays exit under our feet), we are pushed down to the earth. This effect weakens with the square of the distance, and all of Newton's laws are still valid, but the picture is now comprehensible.

Einstein's idea of gravity was this: A body in space causes

the surrounding space-time to curve. The curved space-time in turn tells the mass of the body how to generate gravity and cause space-time to curve. This is not a joke! It is expressed this way in every reference book. So, how far can such an effect reach? Certainly not millions of light years as gravity actually does. My upside down theory explains it this way:

The intensity, or energy content, of gravity rays is disturbed around a body in space as a result of the shielding by this body. Because the shielding effect weakens with the square of the distance, we can conceive curved shells of equal intensity around a body in space. The gravitation towards the body is the same within such a shell of a certain distance (the farther away, the weaker). It can easily be comprehended that this effect of shielding will work over every distance, infinitely in fact. This allows us to see how gravity is caused, while Einstein never explained *how* curved space-time causes gravity.

Because the ratio between surface and volume of an object gets larger when the object gets smaller, it is close to infinite for an atom. Because *an atom is a big body in space* for the narrow gravity rays, they press on the nucleus of an atom with a force beyond comprehension, causing it to be and holding it together. But it will not collapse; a spinning causal energy (from the same source as the rays) applies counter pressure from the inside with an equal force. This energy rotates at the speed of light at an extremely short radius, and this velocity is squared in the formula for centrifugal force. It is the $c^2$ of $E=mc^2$!

Gravity and the so-called strong force are the same in my theory. If the atom is split by any means, then it is the causal energy that gets free as the $c^2$ of $E=mc^2$. Pushing from inside, it causes an explosion of wide range (like an atom bomb). This is possible because the direction of this force is right in the first place – out! The now-theorized pulling strong force of extremely short range, carried by the now-hypothesized gluons can never do that; it has the wrong direction to begin with.

The energy of the gravity rays and the counterforce inside the nucleus of an atom are the causal thought energy of the causal intelligence, God. That is what we are playing with when we explode atomic bombs. Of course, stars do the same all the time, but our use is misuse!

Please note that my theory does not talk about a *number* of *particles*, as theorized by others in the past (That does not work.) My theory has gravity rays that work on every subatomic unit of every atom of a body in space. They can be weakened through shielding, but not become fewer.

If gravity forces an atom together, then we have to ask why atoms do not become unstable when shielded from one side (like all atoms at the surface of the earth). The answer is that the shielding of a planet is insignificant if compared to the tremendous force at the nucleus of an atom. But there is a limit, our sun for example. This body is so big that too much energy is absorbed; the atoms become unstable and fall apart in the form of nuclear fission.

After reading the next section about the speed of light, you will see that the whole contents of this book point at the direction of the famous TOE, the theory of everything.

## Light Propagation

The following is the third of my three new theories. The nature of the universe was first, and gravity was second, but all three belong together. The infinity of the cosmos made the other two possible, but now we will see that gravity and light propagation work together, based on the same condition.

Albert Einstein accepted the so-called constancy of the speed of light as the fact it is, even though it is a gross violation of all known physical laws, especially the concept of relative motion. He did not worry about how this could be, and set to work on a solution. So he took the laws of relative motion by Galileo and added electromagnetic radiation (unknown to Galileo). He did not see that this was a mistake, because electromagnetic radiation and gravity do not belong in relativity. However, in 1905 he had finished his first work and published it under the name Special Relativity. Naturally, the results are strange because, as I said, gravity and light do not belong in relativity. (Except $E=mc^2$, which is a great contribution to quantum physics.)

The whole Special Relativity theory is just one great error in logical reasoning, built on wrong conclusions. So, what is right? We have to see that gravity and light belong in a theory of their own, completely separate from all other physical and astrophysical theories.

*First*, we must find out *why* and *how* the speed of light is always the same, regardless of the frame of reference. *Then*, we can apply logical reasoning and common sense.

If we consider that all frequencies of electromagnetic radiation, including light, always have exactly the same speed, and that every photon in the cosmos can move through space for billions of years, then there is only one possible logical answer: An *under-laying cosmic carrier system* that has the speed of light! – *Cosmic conveyor belts!* I postulate that the carriers of this system are identical to the gravity rays. For this section I will call these gravity/carrier rays just carrier rays.

The photons of every radiation frequency "take a ride" on this rays. And just as a sack of sand dropped on a conveyor belt will instantly move at the speed of the belt, so will a photon instantly move at the speed of the carrier rays, which is the speed of light (all over the cosmos). We already know that these rays pass through a planet with very little loss of energy. The detailed picture looks like this:

When a photon is ejected or reflected from an atom, it jumps on a ray that exits from the atom at that point. After the ride, which can be a fraction of a second or billions of years, the photon gets stuck in the first atom it hits, or gets reflected. But the ray continues on its way through this piece of matter. If the atoms of a material have large open spaces between them, then the photons can stay on the ray, but the ray is subject to the tiny gravity fields of each molecule and the trajectory will be wavy. The speed through this material *appears* to be a little slower because of the longer way. We call these materials transparent (glass, water, etc.). In non-transparent materials the molecules are so close together that the ray must go through them and the photons have to get off.

A black hole is the ultimate solid body. Because it absorbs *all* gravity/carrier rays, no rays exit on the other side and no photons can enter for a ride. The effect is that no light can come from a black hole; it is invisible. The current theory goes that the extreme gravity holds the light back. The real reason is the fact that no carrier rays exit from the surface, and therefore light cannot take a ride away from the black hole. It was this scenario on the surface of a black hole that made it clear to me that the gravity rays and the carrier rays must be the same!

In Chapter X, I listed and explained many reasons why a new theory for the nature of light propagation is needed. I am very sure that my new theory is the answer to the well-known, almost two-centuries-old question: Why is the constancy of the speed of light the most nonsensical thing ever discovered? Or, why is only light behaving so weirdly?

Did I find the solution for the theory of everything? It will take time to be sure, because old paradigms are very hard to overcome, and scientists like to cling to their pet theories, even if they feel that something is wrong with them. The average reader will not see it, but scientists will see what I did right away, and have a hard time accepting it because it is contrary to current "facts", which are in fact only hypotheses. But because everything is based on logical reasoning and nothing is exotic, they will finally have to accept it. And as I said in the introduction: Everything can be easily comprehended, with the only exception of infinity.

The dream of science is to unite relativity and quantum physics. I saw that this is not possible, because while quantum physics deals with reality, relativity is only an excellent exercise in thinking (not logic). You cannot unite fact and fiction into one unit.

Some people will argue that a few of Einstein's predictions proved to be correct. It is true that some of his formulas do work and bring correct results. It took him ten years to make them work because he did not know *what* it was he was working on. It was his picture of nature that was not right. (We could say the same about Newton. He had no idea what gravity is, but all his formulas are correct).

I believe that I am not alone in my opinion about Albert Einstein's theories. First, why are scientists still looking for ways to test and prove his theories after almost a hundred years? Second, Einstein received the Nobel Price for the explanation of the photoelectric effect, but never for his relativity theories. Why? Because nobody is sure they are correct.

What did I unite in order to create a theory of everything? I united science with religion, or more exactly, quantum physics and astronomy with the realm of the causal cosmic intelligence, God. Every topic in this book *is science*!

But even if my theories are not accepted without resistance by scientists, at least I have introduced you to another way of thinking and reasoning about the world and yourself. And if this book has changed your belief system into a system of curiosity and thirst for knowledge, then its purpose is fulfilled.

A Mind once expanded never returns to its original size.
Oliver Wendell Holmes

# APPENDIX I

This and the next appendix are two examples of how you can put the theories in this book to good use when you read an article that ends with the statement that there is no explanation.

In August 1997 we heard in the news that a fountain of antimatter seems to be jetting out from the center of our galaxy. The scientists observed gamma rays that had the specific wavelength of annihilation between electrons and positrons (positrons are the anti-parts of electrons). It was an indirect observation; nothing could be seen.

We also learned that this poses more questions than we have answers for. The greatest question was where the antimatter came from, and how it can exist or be generated in our galaxy of matter.

This huge cloud of annihilation is dense, close to the galaxy center, and less dense farther away.

The theories in this book enable us to find an explanation that makes sense. Here is my explanation:

From a quasar, where the incoming antimatter black hole was larger than our black hole, the surplus antimatter streamed into our universe in the form of a long, stretched cloud of positrons. Now, after a long journey of billions of light years, it got into the gravitational field of our galaxy and accelerates into the center of the Milky Way.

Our galaxy is surrounded by a huge halo of electrons from the solar winds of its over 100 billion stars. When the antimatter cloud of positrons enters this halo, collisions of electrons with positrons occur, and they annihilate each other. Because the electron halo gets more dense towards the center of the galaxy, the frequency of collisions goes up. This picture results in the false impression that the antimatter is a product of our galaxy and is moving outwards. An exotic black hole that disregards our laws of physics, as some scientists theorize, cannot be the source of the antimatter. No, it came from another universe, one composed of antimatter! (See Chapter IX.)

# APPENDIX II

A small, very hot object, billions of light years away, produced one of the most powerful gamma ray bursts ever observed. Within half a year, it grew into a gigantic fireball more than a trillion miles wide, and was still expanding at the speed of light at that point. At the start, in May 1997, the small initial hot point raced outward, away from us, at the speed of light. But after six months, the immense fireball slowed down to 85 per cent of the speed of light.

What is it? It is the most powerful thing ever observed, but beyond that, only speculation exists. Scientists had no explanation. But the theories in this book direct us to a specific explanation that covers all unexplained observations: a huge black hole, leaving our universe at the speed of light, had a head-on collision with a much smaller antimatter black hole that arrived from another (antimatter) universe. It could not stop our big one, but slowed it down.

Only the antimatter black hole, and an equal mass from our large one, annihilated each other. The remaining mass of our matter black hole was completely destroyed by the tremendous energy of the initial quasar and changed into a very fast expanding cloud of hot plasma. Because the mass of the super-compact black hole changed almost instantly into a gaseous form, it required suddenly more space by magnitudes. Theoretically, this immense space should be completely occupied instantly, but because the speed of light is the cosmic speed limit, it can ex-expand "only" at the speed of light. And while expanding, it cooled down, and so did the observed frequency of the radiation. After the initial gamma ray burst, only 15 seconds in this case, it went down through the whole spectrum of electromagnetic radiation – first the X-rays, then the ultra-violet, visible light, infrared, and finally the radio frequencies.

This explanation of the above observations makes sense and can be comprehended but, of course, first my new theory of the nature of the universe has to be accepted, because that is what this explanation is based on.

At the end of 1997, the astronomers at Yale University studied seven black holes in binary systems. The results contradicted all current theories about the nature of black holes. Six of them had nearly equal masses, about seven times the sun's mass, the seventh had about 14 times the mass of our sun. (Could this be two black holes? (My comment).)

It is said that no one has yet been able to explain this surprising result. But I think that I have an explanation, based on my theory of gravity.

According to my theory a super-dense body in space will become a black hole when *all* gravity rays are absorbed. This means that surface gravitation, or G-constant, will have the maximum of gravitation, more is not possible (that would be more than *all*). Therefore, if other black holes have ten or a thousand times more mass, the surface gravity remains always the same (the absolute cosmic maximum). This is so, because the shielding remains the same, the greatest possible maximum. Even the so-called event horizon will always be at the same distance from the surface.

So, what do we have? The masses of the black holes were calculated by their observed gravitational power. According to my theory, they only seemed to be of equal mass. In reality we got this impression because all sizes of black holes have the same surface gravitation. Therefore, we still do not know their mass and never will!

Now we can continue to believe in black holes with the mass of a huge sun, or with the mass of a whole galaxy. But we have to keep in mind that their gravitational effect will be very much the same. Only the difference in diameter will result in a slight difference of the shielding effect. The density is always the maximum it can be.

Isn't it nice to know something others do not know? Now, when you read about a problem in cosmology or physics, you can try to solve it for yourself using the theories in this book.

In closing, I will quote Albert Einstein's question at the end of his book, *The Evolution of Physics* (1938):

> Will the further development be along the line chosen in quantum physics, or is it more likely that new revolutionary ideas will be introduced into physics? Will the road of advance make a sharp turn, as it has so often done in the past?

I believe it will not only be a sharp turn, but it has to be a complete U-turn, as shown in this book.

# APPENDIX III

When Albert Einstein worked on the field equations for his General Relativity, he found out that the universe may be expanding. But this purely mathematical result did not agree with his picture of the universe which was: not expanding forever (open), not contracting (closed), but in steady state (flat). So he invented and added to his equations a cosmological constant. It was an arbitrary constant, only mathematical without any idea what it could be in the reality of nature.

When Edwin Hubble found out that the universe is expanding (in 1929) and had solid proof for this fact with the red shift of the galaxies, Einstein had to change his opinion and called his cosmological constant his greatest blunder.

But now, around the turn of the century, the observing astronomers found proof that the expansion of the universe is accelerating, the farther out the faster. This is a great problem, because it does not agree with the laws of physics if we consider the Big Bang idea – and almost everybody believes in the Big Bang theory. Our cosmologists are working hard to find an explanation. All they have come up with so far is a so-called *dark energy*, which is in reality nothing but a cosmological constant by another name. Nobody has any idea what it could be!

My theory of the nature of the universe (rotating) provides an easy and simple answer: this mysterious energy is simply the centrifugal force of our rotating universe. And because the velocity of rotation is greater the farther out, and because this velocity enters the formula for centrifugal force squared, the expansion must be accelerating. It is the same invisible, mysterious power that forces a speeding truck out of a curve and off the road. The only difference is the scale. For the truck a few feet and a few miles per hour make all the difference, in our universe we have light years and speeds up to the speed of light (at the rim).

My theory does not need a dark energy and no dark mass either. It has no big bang and no inflation. The universe rotates like everything else in the cosmos. It operates by logic, common sense, and by the laws of physics we know and not by exotic mathematics.

# APPENDIX IV

We read and hear a lot about dinosaurs, but almost never that dinosaurs are impossible. Let us think about it.

An African elephant is, at seven tons, the largest land animal and is the upper limit in size for many reasons. Their mating procedure is almost a disaster, but they manage.

If we double all dimensions of such an elephant, the weight will not be doubled, it will be 56 tons, eight times as much because the volume of the animal will be cubed. But the cross section of the muscles will only be squared, or four times as much. They cannot hold and operate these 56 tons; they are too weak for this task. Now, if we consider that the largest dinosaur on record had 80 tons, we can say that such dinosaurs are absolutely impossible. They could never have existed.

Problem is, they did exist and they were walking the Earth on their four legs. I see also a great problem with *Tyrannosaurus Rex* with his ten tons. So far, I have found no explanation for the tiny, almost useless front legs in our literature. Why? I do not see a problem with this. If we compare the design of T-Rex with the design of a kangaroo, then we can clearly see that this ten-ton monster moved along by leaping! But again, we cannot imagine an animal one and a half times the weight of an elephant jumping around. It is also impossible, but they did.

Even worse are the birds. Our largest bird, the albatross with its 30 pounds has a very hard time to take off and the landing is always a comical-looking crash-landing. This bird is the limit for flying in our atmosphere. But what about the remains of the 200 pound Argentinean teratorn from the dinosaur era? It should be impossible, but it did exist and fly!

The only possible explanation for the above problems is an

Earth that had a much lower gravity at that time. But this is, again, impossible because gravity as understood today, is the result of the Earth's mass, and that could never have changed.

How can we deal with this perfect oxymoron? If we consider my theory of gravity which is pushing and based on a shielding effect, then the whole paradox can be solved in a very easy way.

Now let us imagine that our solar system passed through a huge cloud of interstellar gas, a so-called nebula with the mass of a few hundred suns? What happened when our solar system moved through this cloud from about 200 million years ago until 60 million years ago? During this time this nebula would shield our system from all sides and gravitation would be way down. And because the ratios did not change, the solar system kept running like it always did. Only under these conditions, a much lower gravity, dinosaurs of these large dimensions could exist. I believe that this was the reason why the dinosaurs existed at all at that time!

When our solar system entered this huge cloud 200 million years ago, the conditions were terrible. For 12 hours the gravitation was normal, and for 12 hours it was extremely low. The result was disasters of all kinds and almost all life was wiped out. Then, when we were deep within the cloud, the low gravitation was steady and the dinosaurs could be introduced. But 60 million years ago, when our solar system moved out of the cloud, the same terrible conditions as 200 million years ago were repeated. This alternating change from normal gravity to extreme low gravity in a 12 by 12 hour cycle was a problem for the dinosaurs. The extreme low gravity did not bother them, but the 12 hours of normal gravity were too much and they became extinct. But all the small animals could take that and survived.

The above scenario could be a good explanation for the archaeological findings: A terrible wipe-out of most life 200 million years ago and the end of the dinosaurs during a span of about 10 million years that happened about 60 million years ago. The huge meteor that hit near Yucatan at that time could never be so selective. And it was not a dramatic climate change either. It was a gravitational catastrophe!

When we think logically about the whole content of this appendix it becomes clear that only my theory of gravity provides a solution. Gravity as an attracting force, as is understood today, cannot solve this real problem and leaves as with an unsolved oxymoron. I see a good proof for my theory in these facts. It *must* be like that!

# APPENDIX V

## The Ultimate Question

Hundreds of books explain our universe from all possible viewpoints. We learn about the huge sizes of stars and galaxies and about the tremendous distances between galaxies. (Chapters II and III of this book do the same.)

Hollywood and science fiction books try to tell us that the universe is filled with wars and violence and that some day very bad aliens may land on Earth and destroy everything.

Our scientists show us that all of this is practically impossible because the distances between stars and galaxies are forbidding. A trip to our nearest star would require four-and-a-half years at the speed of light, which is impossible for us. And here is my ultimate question: WHY IS THE UNIVERSE THE WAY IT IS?

The universe is filled with life, we are sure about that. But what is the use of such a set-up if the inhabitants cannot communicate or visit? It makes no sense. But we can, as we will see.

The speed of light as an absolute limit for matter and energies exists for a reason. It is a built-in safety feature that prevents ignorant people like us and from other planets from visiting other star systems with destructive weapons on board and the tendency to shoot at everything they do not understand. We cannot bother anyone out there, and all the ignorant people from other planets cannot bother us either.

The physical cosmos is a thought manifestation out of the mind of the Causal Cosmic Intelligence, which we call God. This omnipresent mind *is* everything and is *in* everything! And He set laws, rules, and safety features for His creation. The speed of light as a limit for everything physical and all energies is the most important safety feature.

Now let us talk about the spiritual realm, which is the primary and causal energy. Physical stuff, like our universe, is secondary and only a kind of playground for souls and other spiritual entities.

And here is the clue: Spirit is not limited by the speed of light! Spiritual beings, or 100 per cent spiritual-oriented people can change location instantaneously. They can move from one galaxy to another in zero time. That is the real space travel – and travel is heavy in our universe and all others.

Why is that so? The spiritual realm of the cosmos runs by the law of *love*, which is a force! Our laws are mainly based on the power of one group and enforced on all others. They are inferior and will not work for interstellar connections. We still have a long way to go before we reach the spiritual level that is required for joining the galactic community.

An infinite number of planets in the infinite cosmos has reached this level. There must be hundreds in every galaxy. They have no crime, no wars, only total love and total understanding of the purpose of their existence. They keep their planets natural as much as possible and get their resources from the lifeless planets and moons of their solar system.

What can they do in respect to travel? Most of their travel and transportation is done by what we call teleportation – spiritual power over matter. For space travel they change the ship and the crew into a spiritual entity, using their highly advanced technology which combines spiritual and physical laws. All their ships are mind-controlled. By means of teleportation they change location instantly and move from one solar system to another in no time. At arrival they can remain spiritual and be invisible, or they re-manifest the ship into matter and continue to operate by gravity control.

This is the only way to fly out there. The speed-of-light-safety-feature as a limit for matter or energy prevents us and all other underdeveloped planets from doing the same.

Can we communicate with them by radio waves or light? *No!* They communicate entirely by telepathy, just like we did before we developed language. All our animals communicate in this way; their sounds are only used to express emotions. Many people have already relearned to communicate with animals telepathically. Some teamed up very well with veterinarians.

What is the conclusion? As long as our science is materialistic, as long as we have hate, crime and wars, we are not ready to join the galactic community. The speed of light keeps us within our solar system and we cannot bother anyone else. But for beings of the spirit, or people who live in harmony with spirit and the laws of the universe, there is no limit for travel. The whole universe is theirs for travel in no time!

# SUGGESTED READING

> I cannot live without books.
>
> Thomas Jefferson

## Astronomy

Barbree, Jay and Martin Caidin, *A Journey Through Time*, (Hubble Telescope), New York, Penguin, 1995

Boslough, John, *Masters of Time*, New York, Addison Wesley, 1992

Illingworth, Valery, *The Facts on File Dictionary of Astronomy*, New York, Facts on File, Inc. 1994

Trefil, James, *Space Time Infinity*, Washington D.C., Smithsonian Books, 1985

## General Knowledge

Foster, Dr. David, *The Philosophical Scientists*, New York, Barnes & Noble, 1993

*Reader's Digest*, "Into the Unknown", Pleasantville, 1981

*Reader's Digest*, "The World's Last Mysteries", Pleasantville, 1978

Taylor, Prof. John, *When the Clock Struck Zero*, New York, St. Martin's Press, 1994

## Human History

Cayce, Edgar Evans, *Edgar Cayce on Atlantis*, New York, Paperback Library, 1969

Donnelly, Ignatius, *Atlantis, The Antediluvian World*, New York, Gramercy Publishing, 1949

Gilbert, Adrian G., *The Mayan Prophecies*, Rockport, Mass.,

Element Books, 1995

Robinson, Lytle, *Edgar Cayce's Story of the Origin and Destiny of Man*, New York, Berkley, 1976

## Inspirational

Churchward, James, *The Lost Continent of Mu*, New York, Paperback Library, 1968

Stearn, Jess, *Edgar Cayce, The Sleeping Prophet*, New York, Bantam Books, 1981

Stearn, Jess, *The Door into the Future*, New York, MacFadder-Bartell, 1963

Sugrue, Thomas, *There Is A River*, Virginia Beach, ARE Press, 1993

## Philosophy

Beardsley, Monroe, *The European Philosophers from Descartes to Nietzsche*, New York, Random House, 1960

Kant, Immanuel, *Critique of Pure Reason*, Garden City, New York, Doubleday, 1966

Nietzsche, Friedrich, *Beyond Good and Evil*, Middlesex, Great Britain, Penguin Books, 1984

## Physics

Einstein, Albert, *Relativity*, New York, Crown Publishers, 1961

Zukav, Gary, *The Dancing Wu Li Masters*, New York, Bantam Books, 1986

## Reincarnation and Karma

Cayce, Hugh Lynn, *Edgar Cayce's Story of Karma*, New York, Berkley Books, 1984

Cerminara, Gina, *Many Mansions*, Virginia Beach, ARE Press, 1966

Langley, Noel, *Edgar Cayce on Reincarnation*, New York, Paperback Library, 1969

Martin, Eva, *Reincarnation, The Ring of Return*. New York, University Books, 1964

Shelley, Violet, *Reincarnation*, Virginia Beach, ARE Press, 1986

Stearn, Jess, *Soulmates*, New York, Bantam Books, 1985

## Religion

Arnold, Sir Edwin, *The Song Celestial, or Bhagavad Gita*, London, Great Britain, Routledge & Kegan, 1964

Furst, Jeffrey, *Edgar Cayce's Story of Jesus*, New York, Coward-McCann, 1969

*Revelation, An Edgar Cayce Commentary*, Virginia Beach, ARE Press, 1993

Rinpoche, Sogyal, *The Tibetan Book Of Living And Dying*, San Francisco, Harper, 1992

*The Holy Bible*, King James Version

Yogananda, Paramahansa, *Man's Eternal Quest*, Los Angeles, Self-Realization Fellowship, 1975

Yukteswar, Sri Swami, *The Holy Science*, Bihar, India, Yogoda Society, 1963

## Spirituality

Atwater, P. M. H., Lh.D., *Beyond the Light*, New York, Avon, 1994

Brinkley, Dannion, *Saved By The Light*, New York, Villard Books, 1994

Moody, Raymond, M.D., *Life After Life*, New York, Bantam Books, 1976

Newton, Michael, Ph.D., *Journey of Souls*, St. Paul, Minn., Llewellin, 1996

Pritchett, Blanche, Ph.D., *The Vedas* (Translation), Arlington, Wash., Marcap, 1965

Whitton, Joel, M.D., and Joe Fisher, *Life Between Life*, New York, Doubleday, 1986

Yogananda, Paramahansa, *Autobiography of a Yogi*, Los Angeles, Self-Realization Fellowship, 1994

www.ingramcontent.com/pod-product-compliance
Lightning Source LLC
Chambersburg PA
CBHW020634220526
45464CB00001B/143